The Cybersecurity Endgame

America's Race Against Time

AF881260

By
David Trader

The Cybersecurity Endgame: America's Race Against Time

Copyright © Dave Trader (2024)

All rights reserved. No part of this publication may be reproduced, stored in a retrieval system, or transmitted, in any form or by any means, without the prior written permission of the publisher.

Published by:

AUTHORS on MISSION

TABLE OF CONTENTS

INTRODUCTION

Every day, new threats emerge in the silent corners of our digital existence, unveiling a battleground where freedom, privacy, and the very essence of democracy are constantly under siege. This is not a war fought with bombs and bullets, but with lines of code and stolen credentials. It's a silent conflict raging beneath the surface of our daily lives, a war where the battlefields are invisible and the consequences can be devastating. As patriots, we stand united, resolute in our commitment to safeguarding the values we hold dear. The United States, propelled by rapid technological advancement, once again takes its place at the forefront of an unprecedented battle—a cybersecurity endgame.

In this new era, the stakes are higher than ever, and our adversaries are elusive and sophisticated, posing a unique challenge to our nation. As we navigate the complexities of this digital landscape, we must remember the lessons from past battles, such as the infamous WannaCry ransomware attack that erupted onto the world stage in 2017. Within hours, hospitals plunged into chaos as critical systems were locked down and patient records were held hostage. The reverberations of WannaCry didn't stop at the healthcare sector; they spanned industries and borders, infecting over 200,000 computers in 150 countries, including the United States.

As nations grow more interconnected and the power of technology becomes more accessible, mere ideas can easily become threats to the foundation of our society. The US, once a pioneer in cyberspace, now finds itself locked in a perpetual struggle against state-sponsored hackers, criminal organizations, and lone wolves wielding malicious code. Just like the attack on Pearl Harbor propelled the US into World War II, these digital incursions have thrust nations into a constant state of cyber warfare—a race against time to secure our digital future.

The urgency of this race is undeniable, clearly demonstrated by the persistent occurrence of high-profile cyber incidents that continue to afflict our nation and the organizations within our borders. These incidents underscore the critical need for the United States to intensify its efforts in combating cyber threats and securing its position at the forefront of technological innovation.

On August 3rd, 2023, a major US hospital network fell victim to a ransomware-based cyberattack, causing outrage and disruption. The attack affected sixteen hospitals in multiple states, forcing some to cease operations and transfer patients to other facilities.

A few months before the devastating attack, a storm was already brewing for PharMerica, a major US pharmacy network. Between March 12th and 13th, a data breach ripped through their systems, exposing the private information of a staggering 5.8 million patients. Imagine the shock and anger at names, addresses, birth dates, and Social Security numbers all spilled out. But it wasn't just sensitive identifiers affected. The assailants also accessed details about patients' healthcare providers, medical histories, and even their medication lists. This meant the most intimate aspects of people's lives, their health struggles and vulnerabilities, were now potentially in the hands of criminals.

The financial toll was also significant. PharMerica faced lawsuits, regulatory fines, and reputational damage that threatened the very core of their business. But the true cost wasn't measured in dollars; it was etched in the anxiety and fear of millions of individuals whose personal information was compromised. Imagine the sleepless nights spent worrying about identity theft, fraudulent charges, or even discrimination based on stolen medical data. For these patients, the breach wasn't just a statistic; it was a deeply personal violation with potentially life-altering consequences.

Cyberattacks are becoming a more serious threat for individuals and businesses of all sizes in today's digital world. Cybercriminals are becoming increasingly sophisticated, and their attacks are more costly and destructive. A recent study found that the average cost of a data breach to an organization is $4.45 million, and the cost of cybercrime is expected to reach $10.5 trillion. The FBI revealed that Americans alone have lost more than $4.1 billion to cybercrime, making clear the size of the growing issue.

A few years ago, Colonial Pipeline, a major US fuel pipeline operator that carries 45% of the East Coast's fuel supplies, was forced to shut down its operations after a cyberattack. The attack resulted in widespread fuel shortages and economic losses worth billions of dollars. That same year, the largest meatpacker in the world, JBS, was forced to pay cybercriminals a ransom of $11 million after a cyberattack interrupted its own business.

Not only does a cyberattack impose significant financial costs and operational disruptions on an organization, but it can also have profound personal and professional repercussions for those in leadership positions.

For executives and senior management, a cybersecurity breach can be a career-defining moment. The aftermath may require them to navigate a complex landscape of crisis management, legal challenges, and reputational damage control. Their ability to effectively guide their team through such a crisis can shape perceptions of their competence and leadership skills, both within the organization and in their broader industry.

Professionally, the incident may lead to increased scrutiny from regulatory bodies, stakeholders, and the public. Executives may find themselves answering tough questions about the adequacy of their organization's cybersecurity measures and potentially facing legal consequences for any perceived negligence. Moreover, the lasting impact on the company's reputation may influence career trajectories and opportunities for advancement.

On a personal level, the stress and pressure associated with managing the fallout from a cyberattack can take a toll. The demands of crisis response, coupled with the weight of responsibility, can affect the mental health and well-being of those at the helm. Striking a balance between professional obligations and personal well-being becomes an additional challenge in the wake of such incidents.

Earlier this year, a fraudster called Ryushi demanded $200,000 from Twitter, claiming to have stolen 220 million users' email addresses. He even threatened to sell the data he'd obtained from Twitter to a third party in the event that his demands were not fulfilled. The information that was taken in this incident could easily identify a large number of people, which presents a serious privacy risk. For executives and senior management, this could be very problematic.

But the stakes are no longer only about stealing secrets or disrupting systems. Today, cyberattacks also aim to inflict real-world damage, manipulate elections, cripple economies, and destabilize entire societies. The scars of these digital battles are visible everywhere, from the gas prices sent soaring by the Colonial Pipeline ransomware attack to the hospitals plunged into chaos by the WannaCry compromise. The time for complacency is over. America, as a nation entrusted with safeguarding its citizens' data, must now urgently prioritize cybersecurity in the face of escalating cyber threats.

As we confront the realities of this cybersecurity endgame, we must rise to the challenge and confront the threats that jeopardize the very essence of our democracy and way of life. Just consider the chilling reality: a single attack on a financial institution can trigger not just massive financial losses but also the systematic erosion of consumer trust. Imagine the millions of individuals robbed of their financial security, their credit histories compromised and their savings plundered. The cascading effect can be immense, shaking the very foundations of our economic stability.

But the damage isn't confined to wallets. A cyberattack can also unleash a devastating tsunami of stolen personal data, exposing sensitive medical records, Social Security numbers, and even private communications. The psychological toll on individuals can be immense as the leak leaves them vulnerable to identity theft, discrimination, and even blackmail. The list, unfortunately, goes on. The healthcare sector, critical infrastructure, and even government agencies have all become targets, each breach bringing us closer to a full-blown digital crisis.

Mere acknowledgment of the problem is no longer enough. We need decisive action and a united front from governments, businesses, and citizens alike. Only by investing in robust cyber defenses, fostering international cooperation, and holding perpetrators accountable can we hope to navigate this increasingly digital battlefield and secure a safer future for all.

In 2013, Sony Pictures Entertainment was the victim of a massive cyberattack. After breaking into Sony's computer networks, the hackers took off with a wealth of information that included Social Security numbers, employee salaries, and unreleased movies. They then released the stolen data online and threatened to carry out terrorist attacks on Sony theaters.

The 2013 Sony cyberattack serves as a poignant reminder of the pervasive threat of cyber war. Yet, despite this glaring example, there are areas where the echoes of these ongoing cyber conflicts seem to go unnoticed. One arena where the impact can be particularly catastrophic is the healthcare industry, where cyberattacks not only compromise private data but also pose a direct threat to patients' lives. Healthcare becomes a battleground where the consequences of cyber warfare are felt acutely, with potential implications for public health and safety worldwide.

This painful reality forces us to confront the notion that intellectual property theft is not the endgame of cybercrime; it extends far beyond that, reaching into the lives and well-being of citizens, not only within the borders of a country but also on a global scale.

The Sony crime also demonstrates how susceptible even highly developed companies are to cyberattacks. It shows that cybercriminals are growing more proficient and eager to target any nation or organization, regardless of its size or industry. Imagine that one morning, when your computer screen comes to life, it shows a terrifying message rather than the comforting buzz of productivity. An unknown adversary has taken control of your files, identity, and everything you value. This is not a dystopian novel; rather, it is an example of the real threats that we deal with every day.

These are only a handful of the numerous private incidents and well-publicized cyberattacks that have happened recently. The cyber threat landscape continues to evolve at an alarming pace. This presents a serious risk to individuals, companies, and governments alike. In the US, there were a record 2,116 data compromises between January and September of 2023, exceeding the all-time annual high of 1,862 data compromises set in 2021, according to the Identity Theft Resource Center's (ITRC) third-quarter report.

Over 234 million people, representing a staggering 70.21% of the US population, had their personal information exposed in that period. But that's just the tip of the iceberg. This figure only accounts for publicly known cases, hinting at a far more extensive and insidious undercurrent of cyber threats lurking beneath the surface.

This sobering reality underscores the urgent need for our nation to implement solid cybersecurity measures. The ever-escalating complexity and frequency of cyberattacks demand immediate action. The stakes are simply too high. Hackers, emboldened by their growing sophistication, constantly target individuals and companies, aiming to steal not just money but also their personal data.

The world of cyberspace is changing, so we need to keep up with the times, build effective defenses, and adapt. If we do not take steps to improve our cybersecurity posture, we could face serious consequences. Digital warfare affects America and other governments around the world. In cyberspace, nations rise with one another and with other political entities. Throughout the years, North Korea, China, and other nations have made us their top target. They use cyberattacks as a tactic for waging wars and stealing trade secrets and intellectual property from the US through cyber espionage.

Indeed, cyber warfare has become a potent tool and, in some cases, a precursor to traditional armed hostilities. Its strategic value lies in its ability to cripple an adversary's military capabilities, disrupt essential infrastructure, and undermine public morale. The success of a cyberattack can provide a decisive advantage in a conflict, making it a compelling choice for those seeking the upper hand.

A notable example occurred in the years leading up to the 2022 invasion of Ukraine. Russia orchestrated a series of devastating cyberattacks against the country, targeting critical sectors such as the electricity grid, banks, government offices, and transit networks. These attacks resulted in significant upheaval and financial losses, showcasing the potential impact of cyber operations on a nation's stability and resilience.

From a historical perspective, cyber warfare represents a unique form of conflict, diverging from traditional methods. The battleground is virtual, and adversaries can wage incursions from behind screens, challenging conventional notions of war. As a veteran, I find it particularly striking how the dynamics of warfare have evolved, with technology now playing a central role in shaping the nature of conflicts.

Mounting a defense could become more difficult in the event of a cyberattack on a military network that scrambles coordination and communication. Furthermore, a cyberattack on a social media or government website could sow discord and propagate misinformation among an adversary's populace.

Cyberattacks can also be used to find possible targets for physical attacks and obtain intelligence on an enemy. A physical invasion or other military operation can then be planned and carried out using this intelligence. For instance, in the years leading up to the invasion, Russia utilized its cyberattacks against Ukraine to map out the nation's essential infrastructure and identify vulnerabilities in its military networks. The invasion was then planned and carried out with the help of this intelligence.

It is expected that cyber warfare will increase in frequency as a prelude to conflict as it develops in sophistication and destructiveness. Countries are improving their cybersecurity postures due to growing awareness of the dangers. However, no country is immune to cyberattacks, and we will likely see more of them being used as a prelude to war in the future.

According to the recent Waterfall 2023 Threat Report, there was a 140% increase in cyberattacks against industrial operations in 2022, resulting in over 150 incidents. The report issues a dire warning, stating that at this rate of growth, we can expect cyberattacks to shut down 15,000 industrial sites in 2027—less than five years from now. Here are some other worrying numbers highlighted in the incidence statistics of the Waterfall Report:

- Numerous disruptions were caused at well-known businesses, including 14 of a leading automaker's plants, 23 of a reputable brand's tire factories, and a significant food and publishing company
- Four distinct attacks resulted in flight delays for tens of thousands of passengers
- Four attacks on mining and metal companies affected physical operations; one of these caused material equipment damage and a fire
- Malfunctions were created in the fuel, bulk oil, and cargo container loading and unloading processes at six seaports across three continents

- Two victim organizations went bankrupt as a result of attacks.

While ransomware attacks have financial motives at their core, attacks on the industrial sector also attract hacktivists, whose motivations are ideological or political. There was never a ransom demand in any of the hacktivist attacks in 2022. According to Waterfall, the only goal of each incident was to interfere with vital services or infrastructure.

The conflict between Russia and Ukraine and the ongoing hostilities between Iran and Israel were closely linked to the majority of hacktivist incidents. Out of the six hacktivist attacks, four caused disturbances to transportation services (such as trains, taxis, or public transportation) and one targeted a steel mill, resulting in equipment damage and a fire. The most recent hacktivist attack was directed towards a power utility's EV charging stations.

It is becoming increasingly clear that enhanced cybersecurity is not just an advantage but an imperative in today's world. The Colonial Pipeline ransomware attack and the SolarWinds hack have shown us that our critical infrastructure is vulnerable to hacktivists and that a concerted effort is required to strengthen our cybersecurity posture. The United States finds itself in a race against time as it navigates the complex landscape of emerging threats and rapidly progressing technologies.

The use of cyberattacks by Russia against Ukraine serves as a stark reminder that such incidents transcend borders and pose a threat to all nations. It showcases the potential for cyberattacks to be employed as weapons of mass disruption, capable of destabilizing the global financial system or causing nationwide blackouts. This new era calls for more than just awareness—it necessitates a relentless pursuit of knowledge, adaptation, and innovation in the realm of cybersecurity.

As we stand at this digital precipice, the lessons from past incidents illuminate the path forward, but the ground beneath our feet shifts with each passing moment. The cyber threat landscape is dynamic and ever-evolving, demanding constant vigilance and timely responses. From hackers, governments, and corporations fueled by ambition and malice, to individuals and organizations seeking to protect their privacy and assets, the stakes have never been higher.

"The Cybersecurity Endgame: America's Race Against Time" serves as a beacon, shedding light on the urgent need for action, collaboration, and innovation in the face of escalating threats. This book unravels the intricate web of cyberattacks, their impact on national security, and the critical importance of bolstering our defenses.

As you journey through the pages of this book, you'll be equipped with a comprehensive understanding of contemporary cyber threats and actionable steps to safeguard yourself and your organization. From exploring efficient cybersecurity best practices, guidelines, and emerging concerns like AI-powered attacks and quantum computing threats, to learning from real-world examples such as the Sony Pictures Entertainment breach, this book prepares you to navigate the rapidly changing digital landscape.

The pursuit of robust cybersecurity is not a solo mission; it requires collective action. "The Cybersecurity Endgame" highlights the importance of collaboration between the private and public sectors, exemplified by initiatives like the FBI's CISO Academy. It empowers you with insights into incident response vs. incident command, enabling your organization to effectively navigate and respond to cyber incidents.

As cyber threats continue to evolve, so too must our defenses. "The Cybersecurity Endgame" gives you the knowledge and tools to fortify your organization's cybersecurity posture and face the impending wave of cyberattacks with confidence. United in our pursuit of a safer, more resilient digital future, we can collectively chart a course that safeguards our privacy, our organizations, and the very foundations of our society.

The digital battleground awaits, but with "The Cybersecurity Endgame" in hand, you'll be well-prepared to protect what matters most. Don't wait until it's too late—start your journey today and secure your future in the face of cybersecurity's relentless race against time.

Chapter 1

THE THREAT LANDSCAPE:

What Are We Up Against?

As we navigate our daily lives, a silent, yet burgeoning, threat looms in the shadows of cyberspace: the relentless onslaught of cyberattacks. The United States of America, a nation under siege, faces an ever-expanding and diverse array of adversaries. From organized cybercriminals to ideologically driven hacktivists, each aims to exploit the inherent weaknesses of our digital systems, causing irreparable harm to individuals, corporations, and our nation's critical infrastructure.

To effectively combat this complex threat landscape, we must understand the motivations that fuel our adversaries in their relentless pursuit. For some, financial gain is the primary driving force as they exploit vulnerabilities for personal profit. Others find their motivation in ideological pursuits, aiming to expose perceived injustices or challenge the status quo through disruptive cyber activities. Regardless of their goals, the urgency of addressing the escalating cybersecurity threats cannot be overstated.

In this chapter, we embark on a journey to uncover the various types of cyber threats that plague our digital world daily. We will explore personal cyber infringements and navigate the realm of commercial and enterprise-level attacks. Among the cyber threats that will be scrutinized are identity theft, fraud, ransomware, denial of service attacks, hacktivism, social engineering, man-in-the-middle attacks, catfishing, and financial scamming. By understanding hackers' tactics, targets, and the evolving nature of their attacks, we arm ourselves with the knowledge necessary to devise effective strategies and countermeasures.

Identity theft is a pervasive and insidious threat that targets the very core of us as individuals. As we interact with the digital world, we leave behind a unique digital footprint—a treasure trove for cyber adversaries intent on exploiting our personal information. When personal data such as Social Security numbers, credit card details, and passport numbers are compromised, the consequences can be devastating. Victims of identity theft often face legal and financial repercussions as their stolen information is used for fraudulent purposes, leaving them vulnerable and bearing the burden of reclaiming their own identities.

Beyond the individual level, corporations and enterprises also face an onslaught of cyber threats that can have far-reaching consequences for both their bottom lines and the well-being of their employees and customers. From data breaches and intellectual property theft to disruptions of critical services, these attacks can cripple businesses and erode trust in the digital world.

In the face of such diverse and evolving threats, it becomes imperative for America to lead the charge in the Cybersecurity Endgame. We must foster vigilance, education, and proactive collaboration to fortify our defenses against the invisible enemies that lurk in the shadows of the digital realm. By taking a proactive stance, we not only safeguard ourselves and our organizations but also ensure the resilience of our nation in the race to secure our digital future.

To illustrate the significance of identity theft, consider a real-world example: an employee at a large financial institution inadvertently falls victim to a phishing attack. The adversary, impersonating the IT department, sends an email to the employee telling them to update their login credentials due to an alleged system upgrade. Unaware of the scam, the employee complies and provides sensitive information.

With the employee's credentials in hand, the adversary gains unauthorized access to the company's internal network and manages to infiltrate databases containing valuable customer information. This data includes names, addresses, and Social Security numbers. Leveraging this trove of personal data, the adversary engages in large-scale identity theft—opening fraudulent bank accounts, applying for credit cards, and manipulating financial records within the company. The enterprise only becomes aware of the breach after customers report unauthorized transactions and instances of identity theft.

This scenario emphasizes the gravity of identity theft on an enterprise level, as well as the critical importance of implementing robust cybersecurity measures, training employees to recognize phishing attempts, and maintaining constant vigilance against breaches.

The rise of digital identity theft can be attributed to hackers exploiting vulnerabilities in digital protocols. Cybercriminals, well-versed in the intricacies of the digital landscape, take advantage of these weaknesses through malware infections, phishing scams, and database hacking. The expanding digital world, where countless individuals conduct business and financial transactions online, increases the vulnerability of personal data to cyber intrusions. As identity theft is a significant risk for millions of Americans, there is an urgent need for governments and private institutions to collaborate and intensify their efforts in combating this escalating threat.

Closely linked to identity theft is fraud. While the concept of fraud isn't new, the digital era has ushered in a new frontier for malevolent actors. The internet's vast and interconnected landscape provides cybercriminals with opportunities to evolve their tactics, capitalizing on anonymity and global reach. Modern fraud introduces two emerging threats: deepfake technology and the creation of synthetic identities.

Deepfake technology, a name derived from "deep learning" and "fake," has emerged as a powerful tool for cybercriminals. Utilizing artificial intelligence and machine learning algorithms, deepfakes create realistic impersonations that are challenging to distinguish from genuine content. Manipulated videos, images, or audio recordings can deceive individuals, organizations, and automated systems into believing they are encountering real individuals or events.

The implications of deepfake technology are far-reaching. In the realm of fraud, cybercriminals exploit deepfakes to trick individuals into providing sensitive information, making unauthorized transactions, or tarnishing someone's reputation. For example, a deepfake video could impersonate a high-ranking executive within a company, deceiving employees into following fraudulent instructions or transferring funds to unauthorized accounts. The sophistication of deepfakes makes it increasingly difficult for organizations and individuals to discern between truth and deception.

Another rising concern in fraud is the creation of synthetic identities. By combining genuine and fake data, cybercriminals construct false identities that appear legitimate. These synthetic identities are used for various types of fraud, including credit card fraud, loan fraud, and identity theft. By blending real and fabricated information, cybercriminals bypass security measures, establishing a seemingly credible online presence.

Synthetic identities pose a unique challenge to businesses and individuals. Unlike traditional identity theft, where real personal information is stolen, synthetic identities are entirely fabricated, making detection and prevention more complex. Built gradually, these identities establish creditworthiness and legitimacy, making them even more difficult to distinguish. When used to apply for loans, credit cards, or other financial services, financial institutions may suffer substantial losses, while individuals may struggle to clear their names and recover creditworthiness.

As consumer services migrate online, individuals must remain vigilant against unexpected calls and requests. Cybercriminals often employ social engineering tactics to trick people into revealing confidential information or making impulsive decisions. Exercising caution when receiving unsolicited calls, emails, or messages—especially those involving requests for personal details or financial transactions—is crucial. Verifying the communication's legitimacy through independent means, such as contacting the organization directly through official channels, helps guard against falling victim to such scams.

To combat the emerging threat of deepfakes and synthetic identities, organizations and individuals must take proactive measures. Businesses should invest in advanced fraud detection systems using machine learning algorithms to identify suspicious behavior and patterns. Educating employees and consumers about the risks associated with deepfakes and synthetic identities raises awareness and fosters a culture of vigilance.

Technological advancements provide cybercriminals with new tools and tactics. Staying informed, adopting robust security measures, and maintaining a healthy skepticism enable individuals and organizations to navigate the digital landscape with increased resilience, protecting against the evolving threat of fraud.

Ransomware remains another pressing cyber threat that demands attention. This type of attack involves encrypting critical files or systems within a company or an individual's computer, rendering them inaccessible. The adversary then demands a ransom, often threatening to release private data if the payment is not made. While numerous ransomware incidents have made headlines over the years, a notable example, aside from the Colonial Pipeline attack mentioned in the introduction, is the Garmin ransomware attack. Garmin, a multinational technology company specializing in

GPS products and services, suffered a significant outage after being targeted by the WastedLocker ransomware.

During the attack, Garmin's online services, including its website, customer support, and aviation services, were disrupted. The company was forced to shut down its systems to prevent the further spread of the ransomware, causing temporary service outages for its users worldwide. Although Garmin did not officially disclose whether a ransom was paid, the incident underscored the potential consequences of falling victim to a ransomware attack.

As technology advances, ransomware threats continue to develop and adapt. "Double extortion" is a new variant, adding an extra layer of pressure on victims by not only encrypting their data but also stealing it before initiating the encryption process. This technique holds organizations hostage, both denying access to critical files and threatening to expose sensitive information. The potential fallout from leaked data, reputational damage, legal repercussions, and financial liabilities increases the stakes for victims.

Recent incidents demonstrate criminals targeting various industries, including the healthcare, finance, and government sectors. Attacks exploit vulnerabilities in outdated software, weak network configurations, or human error, stressing the importance of organizations staying updated with security patches and investing in solid cybersecurity measures.

Amid the ever-increasing array of cyber threats, a particularly noteworthy one is the denial of service (DoS) attack. In a DoS attack, an adversary floods the target system with an excessive amount of traffic, causing the server to crash or become unresponsive. By inundating the system with numerous false requests, the adversary exhausts the available bandwidth, rendering the system incapable of serving its intended users. While disconnecting from the digital environment can help prevent such attacks, doing so is often challenging due to our dependence on digital infrastructure.

One variation of this method is a distributed denial of service (DDoS), wherein the adversaries attack the target system using multiple systems. Finding the source of a DDoS can be challenging, since individual users are unaware that their computers are being used for this purpose. To protect against DDoS attacks, businesses should invest in anti-DDoS solutions, such as firewalls and intrusion detection systems. Additionally, users should have a good understanding of the risks associated with using digital devices and be aware of the signs of a potential attack.

As previously stated, the motivation behind cyberattacks can include the desire to compromise public services, harm a company's brand, or obtain access to private data. Political motivations may also lead cybercriminals to target specific people or organizations.

Hacktivism, a form of digital activism, has been used to achieve this over the years. This cyber threat has evolved into attacks on targets deemed worthy of sabotage. Hacktivists engage in political activism and social commentary, or simply raise awareness about certain issues, by hacking into target systems and creating data breaches. Motives for their activities range from political and environmental to social, but they aim to cause disruption and influence outcomes.

The most high-profile hacktivist group in the last decade has been Anonymous, which launched successful attacks against well-known organizations to target corrupt business practices and defend free speech. The group sees itself as utilizing digital means to oppose authoritarianism and preserve liberty.

Hacktivism, along with other cyber threats like social engineering, man-in-the-middle attacks, and catfishing, contributes to the complexity of cybersecurity challenges.

Social engineering involves manipulating individuals into divulging information or performing actions that may compromise security. Cybercriminals rely on psychological tactics to exploit human vulnerabilities, creating backdoor access to organizational systems. In the realm of hacktivism, social engineering can be a powerful tool for gaining unauthorized entry, as end-users unknowingly facilitate the degradation of a system's security.

Another layer of cyber threat complexity is presented by man-in-the-middle attacks. In this scenario, adversaries intercept sensitive data during digital transactions, exploiting vulnerabilities in the communication channel. Even encryption cannot fully protect data when adversaries position themselves between the sending and receiving parties. Users often place undue trust in hyperlinks sent through emails or SMS, making them susceptible to tampering and manipulation.

Catfishing, the creation of fabricated personas to deceive and manipulate individuals, adds yet another dimension to the cyber threat landscape. In the context of social engineering and hacktivism, catfishing schemes often unfold on platforms where personal information is shared openly. Scammers, operating under false identities, exploit emotional and psychological connections to perpetrate fraud.

Scammers have become adept at forging digital documents and identities to carry out their schemes. They have successfully infiltrated financial institutions, defrauding clients and causing billions of dollars in losses. In many cryptocurrency scams, for instance, scammers lure victims by promising unrealistically high returns on investments. By exploiting victims' desire for quick financial gains, they entice unsuspecting individuals to send money to fraudulent companies, often located in questionable jurisdictions.

These cyber threats intertwine, creating a convoluted web of challenges for individuals and organizations alike. The rise of hacktivism, coupled with social engineering, man-in-the-middle attacks, and catfishing, emphasizes the need for comprehensive cybersecurity measures. Vigilance, education, and technological safeguards have become critical components in navigating and mitigating the risks associated with these multifarious dangers.

As we explore the cyber threat landscape, it's crucial to address the evolving nature of our adversaries. They are not static entities; they adapt and innovate, constantly refining techniques to exploit new vulnerabilities.

One growing concern is the rise of nation-state cyberattacks. Governments worldwide recognize the power and potential of cyber warfare, developing advanced capabilities to destabilize critical infrastructure, compromise national security, and steal sensitive information. State-sponsored actors operate with impunity, shielded by diplomatic red tape, which makes attribution a daunting task.

The emergence of the Internet of Things (IoT) introduces new vulnerabilities. With everyday devices connected to the internet, from smart home appliances to industrial control systems, the attack surface expands exponentially. Malicious actors can exploit these interconnected devices as gateways to infiltrate networks, compromise data, and disrupt essential services.

The incipient threat of GenAI should also be considered. GenAI, or the use of artificial intelligence (AI) in generating sophisticated cyberattacks, highlights how innovative cyber adversaries are. As AI technology continues to advance, they can leverage its capabilities to create more intelligent and automated attacks, making it even more challenging to detect and defend against them.

The use of GenAI in cyberattacks creates a new level of complexity and innovation. AI algorithms can be trained to analyze vast amounts of data, identify patterns, and generate highly targeted attack vectors. This enables cyber adversaries to carry out more precise and efficient incursions, evading traditional security measures and increasing their chances of success.

One example of GenAI in action is the use of AI-powered botnets. By leveraging AI algorithms, cybercriminals can create botnets that autonomously identify vulnerable systems, launch coordinated attacks, and adapt their tactics in real time to evade detection. These AI-powered botnets have the potential to conduct large-scale and highly-developed assaults, exploiting vulnerabilities across networks and causing significant damage.

Another use of GenAI lies in the development of advanced evasion techniques, since AI algorithms can analyze network defenses and identify weaknesses or patterns in security measures. This knowledge can be used to create attack techniques that specifically bypass or exploit these vulnerabilities, making it difficult for traditional security systems to detect and prevent the attacks.

Additionally, AI-powered social engineering attacks pose a significant threat. Cybercriminals can use AI algorithms to analyze and mimic human behavior, enabling them to craft highly convincing phishing emails, scam messages, and other social engineering pitfalls. These AI-generated attacks can be personalized, tailored to the specific characteristics and interests of the target, making them even more convincing and challenging to identify as fraudulent.

The growing refinement of GenAI elicits significant challenges for cybersecurity professionals. However, the use of AI is a double-edged sword. Safeguarding against AI-generated attacks requires the development of advanced AI-powered defense systems capable of detecting and addressing them in real time. These can be employed to identify potential cyber threats, analyze network traffic for patterns of malicious behavior, and automate rapid incident response processes. By utilizing AI for cybersecurity purposes, organizations can augment their defenses and stay one step ahead of cyber adversaries.

To effectively address the threat of GenAI, ongoing research and collaboration between the cybersecurity industry, academia, and government entities are crucial. By sharing threat intelligence, developing AI-based defense strategies, and implementing proactive security measures, we can better protect our digital systems from this emerging danger.

As the cyber threat landscape evolves, adapting protections is imperative. This requires the combined efforts of individuals, organizations, and governments. Education and awareness equip individuals with knowledge and skills to identify and alleviate risks. Organizations must invest in robust cybersecurity measures, including vulnerability assessments, network monitoring, and incident response plans. Governments must enhance international cooperation by sharing threat intelligence and establishing legal frameworks to hold cybercriminals accountable.

In summary, this chapter unpacks the landscape of cybersecurity threats for individuals and corporations, highlighting risks like identity theft, fraudulent schemes, ransomware attacks, denial of service attacks (DoS), hacktivism, social engineering, man-in-the-middle attacks, catfishing, financial scams, and the growing threat of GenAI. Cyber adversaries are increasingly innovative, and recognizing these dangers helps us take proactive measures to mitigate them.

Now armed with an understanding of the threats, upcoming chapters will delve deeper, examining tactics employed by cyber adversaries, major cyber incident case studies, and best practices for individuals and organizations to fortify their defenses. Together, we can navigate the treacherous waters of the cyber threat landscape, emerging stronger and more resilient. Stay vigilant, for the battle for cybersecurity requires constant adaptation and unwavering determination.

Chapter 2

THE ATTACK SURFACE:

Who Are We Up Against?

In the quest to protect our national security, it is imperative to identify and understand the various entities that pose a threat in the ever-evolving cybersecurity landscape. This chapter delves into the diverse range of adversaries and their motivations, shedding light on the challenges we face as we strive to safeguard our digital realm.

Some of the most formidable adversaries we confront are state-sponsored hackers, who possess extensive capabilities and unwavering determination. Countries such as Russia, China, Iran, and North Korea are prominent players in this arena, employing cyber tools for a range of activities, including digital espionage and intellectual property theft. The United States, too, maintains a strong presence in cyberspace, utilizing its own cyber capabilities to counter threats and protect national interests.

State-sponsored actors—individuals or groups supported by their respective governments—boast sophisticated arsenals and deep expertise, enabling them to execute complex attacks that could potentially compromise national security. Their advanced skills necessitate a robust and vigilant defense against their relentless efforts to exploit vulnerabilities in our digital infrastructure.

China and Russia, in particular, are known to employ individuals or groups to carry out cyber operations. These state-sponsored actors often have access to significant resources, advanced tools, and extensive training, enabling them to conduct adept and targeted cyberattacks.

China's state-sponsored hacking groups, believed to have ties to the People's Liberation Army (PLA) or other intelligence agencies, have been linked to various cyber espionage campaigns targeting United States organizations. These groups frequently employ advanced persistent threats (APTs) and highly-developed malware to infiltrate networks, exfiltrate data, and conduct long-term surveillance.

Through factions like the PLA, China has been responsible for significant cyber intrusions and intellectual property thefts. Notably, the PLA's 54th Research Institute has been linked to the high-profile Equifax breach, which compromised the personal data of millions of Americans. It is estimated that China has stolen around $1 trillion worth of United States intellectual property since 2009, underscoring the scale of their economic espionage efforts.

To recruit individuals for state-sponsored cyber activities, including espionage and hacking operations, China employs a variety of methods, such as:

- Academic recruitment: Identifying talented individuals within universities and research institutions, particularly those specializing in computer science, engineering, and related fields. These individuals may be approached through academic competitions, scholarships, or research grants that align with the interests of the Chinese government.

- Military recruitment: Selecting candidates from within the ranks of the People's Liberation Army or other security agencies, providing them with specialized training and resources to carry out cyber operations.
- Talent spotting at hacking competitions: Keeping an eye on participants in national and international hacking competitions, where individuals showcase their technical skills and problem-solving abilities. These events can serve as a proving ground for potential recruits with exceptional talent and expertise in cybersecurity.
- Direct recruitment from the cybersecurity industry: Targeting professionals already working in cybersecurity, offering them incentives to join state-sponsored groups or engage in espionage activities. In some cases, these individuals may be pressured or coerced into collaborating with the Chinese government.
- Leveraging underground hacking communities: Monitoring online forums, dark web marketplaces, and other platforms where hackers congregate to share knowledge, tools, and techniques. Individuals who demonstrate exceptional skills or innovative methods may be approached by state-sponsored entities for recruitment purposes.
- Identifying patriotic individuals: Appealing to individuals who have strong nationalistic sentiments and a desire to contribute to their country's

interests. These individuals may be more susceptible to recruitment by the Chinese government, believing that their actions serve a greater purpose.

- Targeting overseas Chinese nationals: Reaching out to Chinese nationals living abroad, using a combination of incentives and patriotic appeals to encourage them to engage in state-sponsored cyber activities. China also engages in a strategy known as transnational repression (TNR), which involves targeting Chinese citizens working or studying abroad, including those within US borders. The Chinese government, often through the PLA or Ministry of State Security, coerces these individuals by threatening them or their families, forcing them to act as assets or spies. This tactic, though less discussed, allows China to place key assets in soft target areas, thus extending its influence and intelligence-gathering capabilities.

It is important to note that the specifics of China's recruitment methods for state-sponsored cyber activities may not be fully disclosed or publicly known. These methods are generally conducted with a high level of secrecy and discretion.

Russia is likewise known for its state-sponsored cyber activities, with groups linked to the Russian intelligence community involved in various online operations. These include large-scale cyber campaigns like the SolarWinds attack, designed to compromise government networks and gather sensitive information. Russia's cyber capabilities also extend to conducting influence operations, spreading disinformation, and undermining critical infrastructure.

Iran, too, has emerged as a significant cyber threat actor. Multiple cyber espionage campaigns and disruptive attacks have been attributed to Iranian hacker groups such as Advanced Persistent Threat 33 (APT33) and APT34. These actors have targeted government agencies, critical infrastructure, and industries including the energy, telecommunications, and financial sectors.

South Korea, while a target of cyber threats itself, has also been associated with state-sponsored hacking activities. Notably, North Korea's Lazarus Group, believed to be linked to the North Korean government, has carried out cyberattacks including the infamous 2014 Sony Pictures hack and various financial sector attacks aimed at generating funds for the regime.

These factions and their activities highlight the diverse cybersecurity risks we face and why it is crucial to continue strengthening defenses, fostering intelligence sharing, and promoting international collaboration to mitigate their threat.

State-sponsored actors from both China and Russia are often tasked with gathering valuable intellectual property, compromising government networks, conducting political and economic espionage, and exerting influence on the global stage. They leverage various methods, including malware, social engineering, and zero-day vulnerabilities, to accomplish their objectives.

Beyond the state actors, a secondary threat emanates from non-state entities—hacktivists and cybercriminals. Hacktivists, driven by ideological fervor, employ hacking as a means to advance their political or social causes, often resorting to disruptive and destructive maneuvers. In stark contrast, cybercriminals are motivated by the allure of financial gain, deploying tactics like ransomware, phishing, and identity theft to pilfer valuable data.

While external actors pose significant risks, the potential for threats also lies within our own organizations. Insider threats, whether stemming from malicious intent or simple negligence, can jeopardize our security. Thus, it's crucial to implement strong access controls and maintain vigilant monitoring to identify and mitigate any harmful activities originating from internal sources.

Within this expansive theater of cyber warfare, where conflicts unfold in silence and without physical confrontation, China and Russia continue to emerge as the chief adversaries confronting our country. The sheer magnitude of the threat they pose is staggering, with China alone orchestrating a daily barrage of attacks surpassing those launched by all other nations combined.

China, with its formidable resources and unwavering resolve, stands as a powerhouse in the realm of cyber warfare. Their well-drilled cyber armies, fortified by advanced support systems, execute precise and stealthy maneuvers. From infiltrating government networks to crippling important infrastructure, China's cyber warriors leave no stone unturned in their relentless pursuit of dominance.

Russia's modus operandi, meanwhile, combines state-sponsored activities with those of non-state actors. Their tactics range from disinformation campaigns to disruptive attacks on critical systems, all aimed at sowing chaos and undermining trust in democratic institutions. Masters of deception and manipulation, Russian forces employ a wide array of tactics to achieve their malevolent objectives.

The evolving nature of cyber threats transcends the traditional image of isolated hackers operating from dark basements. Instead, we confront well-organized criminal enterprises with defined structures, quality assurance departments, and even help desks, further complicating the battleground.

The term "threat surface" encapsulates the specific ways in which a threat exploits weaknesses in a system or network to inflict harm. Each threat surface, wielded by hackers, state-sponsored actors, cybercriminals, hacktivist groups, and others, possesses distinct characteristics and capabilities.

As cyber threats progress, they necessitate a more comprehensive and proactive approach to safeguarding our digital lives. Their diversity and constant evolution are products of cybercriminals growing more sophisticated by the day. A profound understanding of the adversaries we face becomes paramount, and exploration of the primary threat surfaces unveils their motivations, techniques, and potential impact on organizations and governments.

It becomes evident that each threat demands our utmost attention and vigilance. One prominent threat surface that looms large in this digital battleground is malware—malicious software designed to infiltrate systems and wreak havoc. From traditional viruses propagating through infected files to high-tech ransomware holding valuable data hostage, malware remains a persistent and formidable adversary.

To better understand the complexities of cybersecurity and the attack surface, it is essential to distinguish between hardware and software, especially for those with limited IT knowledge. Hardware refers to the physical components of a computer or communication system, such as processors, memory chips, circuit boards, and networking equipment. These tangible elements provide the foundation for data processing and communication. Software refers to the intangible components, including the operating system, applications, and programs that run on the hardware. The software allows users to interact with and control the hardware in order to perform various tasks and functions.

In the context of cybersecurity, securing both hardware and software is critical to protecting sensitive information and maintaining the overall integrity of a communication network. Hardware vulnerabilities may arise from physical tampering, design flaws, or the use of counterfeit components that can compromise the system's soundness. On the other hand, software vulnerabilities often result from coding errors, unpatched security flaws, or malicious programs designed to exploit weaknesses in the system.

Whether driven by financial gain or destructive intentions, the creators of malware constantly refine their techniques to exploit cracks in our networks. Malware is as a significant danger weaponized by various countries, including China, to target the United States. This use of malware as a national threat falls under the category of cyber espionage and cyber warfare.

China, with its history of executing sophisticated and persistent cyberattacks against American organizations—both private and governmental—employs diverse malware variants to exploit vulnerabilities in computer systems. Advanced persistent threats (APTs), characterized by their long-term and stealthy nature, are a favorite tool in China's cyber arsenal. These APTs often involve malware such as Trojans, keyloggers, rootkits, and remote access tools.

A noteworthy example of Chinese malware is Advanced Persistent Threat 1 (APT1), also known as Unit 61398, which was exposed by Mandiant in 2013. This APT, linked to the Chinese People's Liberation Army (PLA), was accused of numerous targeted cyber espionage campaigns against our industries. APT1 utilized custom malware tools and social engineering techniques to compromise networks and harvest confidential data.

Operation Aurora, a cyberattack believed to be orchestrated by Chinese actors in 2009, serves as another notable example. This attack targeted multiple US companies, including Google, Adobe, Juniper Networks, Rackspace, and Yahoo, among others. The primary objective of its creators was to gain access to valuable intellectual property and sensitive information. By compromising these companies, the perpetrators aimed to further their economic and strategic interests, highlighting the significance of cyber espionage and the potential consequences for targeted organizations.

China is also known to exploit supply chain vulnerabilities by injecting malware into legitimate software updates or hardware components. This tactic compromises the integrity of the entire system, as witnessed in attacks targeting our government agencies and defense contractors.

The motivations behind China's use of malware as a threat against the US are multifaceted. The objectives include gaining a competitive advantage by stealing intellectual property, acquiring technological advancements, and reducing reliance on foreign technologies. Furthermore, cyber espionage conducted through malware helps China gather political, military, and economic intelligence for its strategic uses.

China has been known to employ social engineering as a significant threat vector against the United States, capitalizing on human psychology to bypass traditional security measures. By exploiting trust and leveraging manipulation tactics, social engineers can deceive individuals into revealing classified information or granting unauthorized access.

A common social engineering tactic employed by Chinese actors is spear phishing, a highly targeted email-based attack that uses carefully crafted messages to lure victims into clicking malicious links or attachments. These attacks often involve extensive research on the target's background, interests, and connections, enabling the adversary to create convincing and personalized messages that increase the likelihood of success.

Chinese threat actors also exploit social media platforms to conduct coordinated influence operations. These operations involve spreading disinformation and propaganda and engaging in astroturfing (creating fake grassroots movements) to manipulate public opinion, influence political discourse, and sow discord within the United States.

These social engineering tactics are often intertwined with malware attacks to amplify their effectiveness. For instance, a phishing email may contain a malicious attachment or a link leading to a website hosting malware. This dual approach significantly heightens the probability of compromising targeted individuals and gaining access to private information.

Russia, too, plays a significant role in espionage and cyber warfare against the United States. Russian hackers and state-sponsored groups stand implicated in numerous high-profile cyberattacks targeting United States institutions, government agencies, and critical infrastructure. Like China, Russia deploys both sophisticated malware and social engineering tactics to achieve its objectives.

In terms of malware, Russia has developed and deployed advanced cyber tools to compromise systems and networks. One notable example is the NotPetya malware, which wreaked global havoc in 2017. Although initially disguised as ransomware, NotPetya was later widely believed to be a Russian cyber weapon primarily targeting Ukraine. Its rapid spread affected numerous organizations worldwide, causing significant disruption and financial losses.

Russian threat actors particularly leverage social engineering techniques like phishing and targeted deception campaigns. They have a track record of utilizing spear phishing emails to gain unauthorized access to sensitive networks and steal valuable data. This often involves the impersonation of trusted entities or the use of persuasive tactics to deceive victims into taking actions that compromise their security.

In the wake of the 2016 US presidential election, concerns about foreign interference in the democratic process have persisted. With another election year approaching, it's important to remain vigilant regarding the ongoing threat of Russian disinformation campaigns and their potential influence on public opinion. These campaigns often rely on social media platforms to disseminate misleading information, exploit existing social divisions, and create confusion among the electorate.

Russia's cyber activities against the US are driven by a range of motivations, including geopolitical interests, intelligence gathering, economic gains, and exerting power internationally. In some instances, the goal is to cause disruption, foment chaos, and weaken confidence in democratic institutions. By undermining the stability and credibility of the US political system, Russia aims to enhance its own position on the global stage and potentially influence decision-making processes in favor of its interests.

By understanding the motivations behind Russia's cyber activities, the United States can better anticipate and defend against potential threats. This knowledge can also inform diplomatic efforts and help guide strategies for mitigating the risks posed by Russian interference, ultimately contributing to a more secure and resilient cyber environment.

To address this issue, various stakeholders, including government agencies, social media platforms, and civil society organizations, have taken steps to enhance election security, combat the spread of false information, and educate the public about potential manipulation tactics. Despite these efforts, the evolving nature of disinformation campaigns and the complexity of the digital landscape require ongoing attention and collaboration to protect the integrity of the electoral process.

As we reach the upcoming election year, it is essential to remain informed about the potential risks associated with foreign interference, particularly social engineering and disinformation agendas. Staying vigilant and promoting media literacy can help individuals identify and counter fabricated information, contributing to a stronger, more secure democratic process.

Zero-day vulnerabilities are another critical threat frequently employed by cyber adversaries against the United States. These vulnerabilities are software flaws or weaknesses unknown to the vendor, with no available patch or fix. Hackers exploit them to gain unauthorized access, steal sensitive data, or facilitate other malicious activities.

Zero-day vulnerabilities have been utilized by state-sponsored actors, including those linked to China and Russia, to bypass security defenses and infiltrate networks. These weak points allow the delivery of malware and the establishment of persistent threats within compromised systems.

The threat posed by zero-day vulnerabilities highlights the importance of robust vulnerability management, prompt patching, and collaboration between software vendors, security researchers, and government entities. Initiatives like bug bounty programs encourage ethical hackers to discover and disclose these weak points, enabling timely patches and reducing the risk of exploitation.

A notable example of a zero-day vulnerability being exploited by Russia against the United States was the SolarWinds cyberattack in late 2020. This sophisticated supply chain attack compromised SolarWinds, a software provider used by government agencies and businesses. It utilized a zero-day vulnerability to gain initial access to the networks of SolarWinds customers and conduct espionage activities, impacting national security and raising concerns over the potential theft of classified information.

This incident illustrated the need for rapid detection and response capabilities, as well as strong supply chain security practices. However, it also underscored the threat posed by state-sponsored actors leveraging zero-day vulnerabilities to carry out large-scale operations against the US and its allies. It shows the importance of collaboration between government agencies, the private sector, and security researchers in identifying and patching vulnerabilities, as well as sharing threat intelligence to prevent similar incidents in the future.

We must also acknowledge the dangers posed by advanced persistent threats (APTs). These are sophisticated and well-resourced adversaries, often state-sponsored or organized criminal enterprises, who engage in prolonged and targeted cyber campaigns with the objective of compromising specific organizations or entities. APTs employ a combination of advanced tools, techniques, and procedures to achieve their goals while remaining undetected.

Chinese and Russian state-sponsored actors, among others, are known for their use of APTs in cyber operations. These adversaries invest significant resources in developing custom malware, conducting extensive reconnaissance, and employing stealthy tactics to infiltrate and persist within target networks.

One notorious Advanced Persistent Threat (APT) group linked to China is APT41, also known as Double Dragon. APT41 stands out due to its engagement in both state-sponsored cyber espionage and financially motivated cybercrime. This duality enables the group to utilize its extensive capabilities to pursue a variety of objectives.

APT41 has been involved in cyber espionage campaigns targeting multiple sectors, including technology, healthcare, and telecommunications. Their highly-refined techniques include the use of custom malware, zero-day exploits, and supply chain attacks to compromise high-value targets. Simultaneously, APT41 has conducted financially motivated cybercrime, such as ransomware attacks and the theft of sensitive information for profit.

Russian APT groups, like Cozy Bear (APT29) and Fancy Bear (APT28), have gained similar notoriety for their cyber operations. These groups are also associated with state-sponsored activities, including cyber espionage and interference in geopolitical affairs.

Cozy Bear and Fancy Bear were implicated in the cyberattacks on the Democratic National Committee

(DNC) during the 2016 US presidential election. These APTs employed various tactics, including spear phishing, malware deployment, and data exfiltration, to achieve their objectives and influence the political landscape.

To defend against APTs, organizations must implement reliable cybersecurity measures, including network segmentation, intrusion detection and prevention systems, and continuous monitoring. Regular security audits, threat intelligence sharing, and incident response planning are other essential components of an effective defense strategy against APTs.

As we navigate the field of cyber threats, it's crucial to recognize the significance of supply chain vulnerabilities as a threat surface. Cyber adversaries often exploit weaknesses in the supply chain to compromise the integrity of software, hardware, or services, thereby gaining unauthorized access to target systems.

The previously discussed SolarWinds cyberattack serves as a glaring example of a supply chain compromise. By infiltrating the software supply chain, threat actors were able to distribute a compromised software update to a vast number of organizations, leading to widespread network intrusions.

In order to mitigate the risks associated with supply chain vulnerabilities, organizations should adopt best practices, including rigorously vetting suppliers, implementing secure development processes, and regularly auditing and monitoring the supply chain for potential risks. Collaboration between industry stakeholders, information sharing, and adherence to established cybersecurity frameworks are also crucial in enhancing supply chain security. More on these will be explained in the next chapter.

In conclusion, our exploration of threat surfaces highlights the diverse and evolving nature of cyber threats faced by the United States. From state-sponsored actors employing advanced methods to sophisticated APTs conducting prolonged cyber campaigns, each hazard poses unique challenges that demand proactive and adaptive cybersecurity strategies.

Understanding the motivations, techniques, and tactics of adversaries such as China and Russia is paramount in developing effective defense mechanisms. The interplay of threat surfaces, including malware, social engineering, zero-day vulnerabilities, and APTs, makes clear the need for a comprehensive and collaborative approach to cybersecurity.

As we confront these challenges, it's essential to remain vigilant, stay informed about new threats, and continually enhance our cybersecurity capabilities. The cyber landscape is dynamic, and our ability to adapt and innovate in the face of its ongoing dangers will determine our success in safeguarding national security and preserving the integrity of our digital ecosystems.

Chapter 3

THE INTERNET AS A VITAL UTILITY:

Prioritizing Security in the Digital Age

Envision a world devoid of the convenience offered by online maps—a world where we're forced to navigate unfamiliar cities relying solely on outdated paper maps or the kindness of strangers for directions. Gone are the days of meticulously planning our routes with physical maps, only to end up making wrong turns and feeling frustrated. Similarly, acquiring information once required trips to the library, countless hours spent flipping through pages, and reliance on cataloging systems and the knowledge of librarians.

The emergence of the internet transformed these experiences, revolutionizing communication, reshaping industries, and unlocking new possibilities. Online shopping, streaming services, remote work, and online education have all become integral parts of our daily lives, ushering in unparalleled convenience, efficiency, and a world teeming with boundless opportunities.

Yet, within this digital world lurks a significant problem—malicious actors and the threats they pose, as revealed in the previous chapters. This compels us to confront a pressing question: is the security of the internet receiving the attention it deserves? As our dependence on the internet grows rapidly, it becomes increasingly important to prioritize its safety. It has woven itself into the very fabric of our lives, presenting countless benefits but also exposing us to various dangers and adversaries.

This chapter delves into the challenges of securing the internet as a utility and the crucial measures required to shield the nation from these adversaries.

To effectively address the evolving cyber threat landscape, a paradigm shift is imperative. We must acknowledge the problems with the internet and begin treating it as an indispensable utility, deserving the same level of protection as water and electricity. Just as these are essential resources that support various aspects of our daily lives, the internet has become an vital tool for communication, information access, and conducting business. It underpins critical functions in sectors such as banking, healthcare, transportation, and emergency services.

Drawing parallels between the internet and traditional utilities like water and electricity enhances our understanding of its need for similar defenses. By viewing them as on par with one another, we underscore the importance of robust cybersecurity measures and regulatory oversight. This entails establishing regulatory bodies to oversee security practices, define security standards, and enforce compliance, ultimately fostering transparency and trust among stakeholders.

The importance of the internet as a utility goes beyond convenience; it's about safeguarding vital services that have a direct impact on our daily lives. In the banking sector, online transactions and access to accounts have become the norm. Healthcare has evolved to include telemedicine, enabling remote consultations, while transportation now relies on ride-hailing services and GPS navigation apps. The internet's profound influence on how we bank, access healthcare, navigate transportation, and communicate with others illustrates its integral role in our lives. Disruptions or compromises in internet security can have far-reaching consequences, emphasizing the need to protect its integrity for the well-being of society.

Drawing from my extensive experience as a cybersecurity expert and former Chief Information Security Officer (CISO), it is clear that conventional cybersecurity approaches are insufficient for addressing the dynamic nature of cyber threats. A shift in perspective is necessary—an acknowledgment that the internet deserves the same level of protection as other critical resources.

This shift requires a proactive and holistic approach to cybersecurity, as relying solely on static defenses and reactive measures is inadequate. The WannaCry ransomware attack is a reminder of the need for preemptive measures like regular patching and vulnerability management to stay ahead of cyber threats.

Moreover, the evolving cyber threat landscape necessitates an adaptive approach that can quickly identify and respond to new hazards. Advanced techniques such as behavior analytics, machine learning, and artificial intelligence can play a pivotal role in detecting and mitigating unknown threats, zero-day exploits, and anomalies.

The concept of digital hygiene is as crucial for online well-being as physical hygiene is for our health. Strengthening password management, device security practices, online awareness, and mindfulness regarding social media engagement can greatly increase our online security.

To further illustrate the gravity of cyber threats, let's look at the Ukrainian power grid attack that occurred in 2015. This cyberattack, believed to be the first known successful attack on a nation's power grid, resulted in widespread outages, affecting approximately 230,000 people in Ukraine.

The adversaries used a combination of malware, known as BlackEnergy, and other tools to infiltrate the networks of Ukrainian power distribution companies. Once inside, they gained control of vital systems and initiated a synchronized assault, disconnecting electrical substations and leaving many homes and businesses without power.

This incident demonstrated the potential for cyberattacks to cause real-world harm and impede essential services like electricity. It marked a significant turning point in cybersecurity, as it showcased the potential consequences of a successful cyberattack on a nation's critical infrastructure.

The Ukrainian power grid attack underscores the importance of securing critical infrastructure against cyber threats and creating robust incident response plans to ensure swift recovery and minimize service disruption. It also highlights the geopolitical implications of cyber threats and the need for international cooperation in addressing the growing challenge of cybersecurity.

This real-world example offers a compelling call to action that cyberattacks can have tangible consequences beyond the digital realm, affecting people's daily lives and the functioning of society. It emphasizes the need for a comprehensive cybersecurity approach, encompassing both prevention and response, as well as collaboration between governments, businesses, and cybersecurity experts to defend our infrastructure and ensure resilience in the face of evolving threats.

Safeguarding the internet as a utility is a matter of national security, but it's also an economic imperative. The stability of American businesses and financial institutions is directly tied to the security of the digital infrastructure on which they rely.

As the United States and other countries around the world explore the implementation of a utility model for cybersecurity, it's crucial to recognize the diversity of regulatory approaches to this issue. In the United States, the approach has generally been more market-driven, with an emphasis on encouraging innovation and allowing industry to drive cybersecurity solutions. Conversely, the European Union has adopted a more stringent regulatory framework with the General Data Protection Regulation (GDPR), which places greater emphasis on individual privacy rights and data protection. Other countries, such as China, follow a state-driven model, with a higher degree of government control over digital infrastructure and data flows.

These differing perspectives on cybersecurity and privacy can pose challenges when attempting to establish a cohesive, global utility model. However, acknowledging and understanding these differences can foster international dialogue and collaboration. This collaboration will ultimately be necessary to overcome obstacles and ensure a safe digital environment that respects individual rights and freedoms while maintaining robust defenses against cyber threats.

As we enter an election year, these issues may become increasingly prominent in public discourse. By recognizing the nuances and complexities of various regulatory approaches, we can facilitate a productive, nonpartisan conversation that focuses on the shared goal of securing the digital landscape for all.

Consider the tension between user privacy and the need for strong cybersecurity. Implementing comprehensive security measures often includes collecting and analyzing user data, which raises legitimate concerns about privacy infringement. Navigating this challenge involves adopting approaches like differential privacy that allow for effective threat detection without compromising individual rights.

To tackle regulatory challenges and privacy concerns, a delicate balance must be struck. One way to achieve this is by incorporating privacy-enhancing technologies, such as zero-knowledge proofs and encrypted data processing techniques, into cybersecurity infrastructure. These methods offer high-quality threat identification while simultaneously preserving the confidentiality of user data.

Zero-knowledge proofs allow one party to prove the validity of a statement to another party without revealing any additional information. This enables verification and authentication processes without exposing sensitive data, making it an essential component of privacy-focused cybersecurity frameworks.

Similarly, encrypted data processing techniques, such as homomorphic encryption, allow computations to be performed on encrypted data without the need for decryption. By processing data in its encrypted form, organizations can analyze and identify threats without accessing or exposing the underlying information, thus safeguarding user privacy.

Additionally, a tiered regulatory approach can be explored, tailoring security requirements based on the criticality of the services provided. Using this strategy, critical infrastructure sectors like energy and healthcare may be subject to more stringent regulations, while less important services face proportionally lighter ones.

As we embark on this paradigm shift, specific and actionable recommendations will be necessary for policymakers, CISOs, and other stakeholders. For policymakers, establishing regulatory bodies with cybersecurity expertise and fostering international cooperation is paramount. Clear, enforceable standards and incentives for compliance can propel the adoption of the utility model.

CISOs and cybersecurity professionals will also play a pivotal role in implementing this shift. Proactive measures, such as regular vulnerability assessments and the adoption of emerging technologies like artificial intelligence, should be at the forefront of organizational strategies. Joint actions with international counterparts and participation in threat intelligence-sharing initiatives can likewise enhance collective cybersecurity efforts.

Public-private collaboration is key to success in this endeavor as well. Encouraging partnerships between government agencies, private sector entities, and academia fosters a symbiotic approach to cybersecurity. Initiatives like the Cybersecurity Tech Accord and cooperative platforms for information sharing create a united front against cyber threats.

Cybersecurity education will be another cornerstone of our defense strategy as we continue to navigate the complex digital landscape. Advocating for its inclusion in American schools and universities is of utmost importance.

Existing educational initiatives and programs aim to improve cybersecurity awareness and skills among students and professionals. Organizations like the Cybersecurity Education and Awareness branch of the Department of Homeland Security work to provide resources, training, and guidance for educators and students. Collaborative steps between academia, industry, and government entities are vital for nurturing the next generation of cybersecurity professionals.

It is worth noting that the United States has recognized the importance of national cybersecurity and has begun implementing measures to strengthen it. Recent government initiatives and legislation reflect a commitment to enhancing our online defenses.

The Cybersecurity and Infrastructure Security Agency (CISA) plays a central role in coordinating these national efforts. The National Cyber Strategy outlines a comprehensive approach to guarding American cyberspace, focused on protecting critical infrastructure, promoting innovation, and fostering a secure digital economy.

Meanwhile, legislative endeavors such as the Cybersecurity Information Sharing Act (CISA) support the sharing of threat information between the government and private sector entities. These initiatives contribute to a more coordinated cybersecurity landscape.

Successful public-private collaborations have demonstrated the effectiveness of a united front against cyber threats. The Cybersecurity Tech Accord, an agreement among major technology companies, exemplifies private sector commitment to cybersecurity. So does the Financial Services Information Sharing and Analysis Center (FS-ISAC), which facilitates the sharing of cybersecurity threat intelligence among financial institutions. By sharing information and resources, these companies enhance their collective defenses and enable a proactive response to emerging threats. Such alliances epitomize the synergy needed between government and private sector entities in bolstering the nation's cyber defenses.

International cooperation on cybersecurity is a necessity. The interconnected nature of the online world transcends national borders, requiring collaborative efforts to combat cyber threats. Partnerships with allies, joint initiatives, and global efforts contribute to a more secure digital environment.

Organizations like INTERPOL, the NATO Cooperative Cyber Defence Centre of Excellence (CCDCOE), and the European Union Agency for Cybersecurity (ENISA) foster international collaboration. These entities strengthen global cybercrime investigations, enhance cyber defense capabilities, and promote a coordinated response to cyber threats.

But internet security is not solely of national interest; it's also deeply intertwined with individual privacy and well-being. The connection between internet security and the everyday lives of Americans is profound. Online privacy, personal data protection, and the implications of cyber threats for individual users are pressing concerns.

Consider the pervasiveness of social media platforms in our lives. The Cambridge Analytica scandal highlighted the misuse of personal data for political purposes, raising awareness about the importance of protecting user information.

To date, numerous organizations and government agencies have successfully implemented reliable cybersecurity measures, preventing significant cyber incidents or minimizing their impact.

The United States Department of Defense (DoD) exemplifies a proactive approach to cybersecurity. Through initiatives like the Defense Industrial Base (DIB) Cybersecurity Program, the DoD partners with defense contractors to enhance the cybersecurity posture of the defense industrial base.

New technologies are shaping the future of cybersecurity. Artificial intelligence, machine learning, and blockchain offer innovative solutions for threat detection, data protection, and secure communication. Integrating these technologies into cybersecurity frameworks enhances our ability to adapt to evolving cyber threats.

Despite our best efforts, major cyber incidents may still occur. Addressing strategies for crisis response is essential to enhance national resilience. Incident response planning, coordination, and teamwork during a cyber crisis are critical components of a robust cybersecurity strategy.

Organizations like the Multi-State Information Sharing and Analysis Center (MS-ISAC) provide resources and support for incident response. Collaboration platforms and exercises expedite coordination between government and private sector entities, ensuring a swift and effective response to cyber incidents.

Ultimately, securing the internet from known adversaries and treating it as a utility is a bold but necessary step in safeguarding our nation's digital landscape. As a cybersecurity influencer, your voice can provoke this paradigm shift as we work together to solve the internet's problems.

By addressing the specific areas highlighted, this chapter strives to provide a more comprehensive coverage of internet security, emphasizing its interconnectedness with national interests, economic well-being, education, legislative measures, collaboration, international cooperation, and the everyday lives of Americans.

Let's work towards a future where the internet is resilient and safe, serving as a foundation for the digital infrastructure on which our society thrives. Your active participation in this paradigm shift can make a significant difference in ensuring a secure and reliable cyberspace for generations to come.

Chapter 4

THE WAY OF WATER

As we embark on our mission to strengthen the internet and address its challenges, let's draw a comparison to a vital resource we often overlook: water. Just as utility providers maintain systems that ensure our access to clean water, the internet depends on a complex infrastructure managed by internet service providers (ISPs). While we've come to expect an uninterrupted flow of data, the investment in securing these internet systems pales in comparison to the comprehensive measures implemented in the water sector.

This chapter introduces the "Way of Water" strategy, an approach that emphasizes the importance of control, architecture, and intake valves in safeguarding our online landscape. By examining the substantial measures taken to protect our water supply, we can glean valuable insights and apply these lessons to fortify the internet against the ever-evolving threats that lurk there. Through this exploration, we aim to inspire a shift in perspective and prompt a renewed commitment to securing the digital realm for the benefit of all users.

Metaphorically, the internet and data networks function much like plumbing. Cables and switches are analogous to pipes and valves controlling data flow, with ISPs acting as the conduits delivering the "water" of data to our homes and businesses. We expect this flow to be clean and safe, just as we expect our water supply to be filtered and treated for impurities.

However, the responsibility for securing this flow of data often falls on individuals and organizations, placing an unfair burden on those who lack the advanced education, expertise and resources needed for effective protection. Unlike water utilities, where consumers aren't tasked with ensuring the purity of their supply, ISPs tend to wash their hands of securing the content passing through their infrastructure.

Imagine if water utilities declared that consumers must now take full responsibility for ensuring the clean and safe delivery of their water supply at home—an outrageous idea. Yet, this is essentially what we tolerate when it comes to our supply of internet access. Despite having the ability to detect and mitigate threats at intake points, powerful ISPs have avoided accountability in addressing the rising contamination of traffic flows.

Major ISPs in the US internet market, such as Comcast, AT&T and Verizon (CATV), possess the power to filter out malicious traffic. However, proactive measures against emerging threats don't align with their financial interests. Testimonies before Congress highlight the reluctance of ISPs to invest in proactive security measures, which they perceive as expenses rather than income streams.

Consider the lack of nationwide competition among broadband providers. In most areas, there are just one or two real high-speed choices, which diminishes the pressure on ISPs to prioritize cybersecurity. This deficiency in consumer choice allows them to focus on profit margins rather than addressing concerns about service quality and transparency.

In the absence of market forces compelling ISPs to prioritize safety, regulatory oversight becomes crucial. As with utilities like water and electricity, where regulations were implemented to safeguard public interests, lawmakers and organizations like the Federal Communications Commission may need to intervene.

Historically, when corporate interests conflicted with public health and safety, society benefited from laws and regulations limiting corporate power. The internet, integral to modern life, may require similar regulatory measures to alleviate the growing risks.

As an example of how ISP oversight is related to critical utilities, imagine a situation in which regulatory dysfunction prevents local water suppliers and treatment facilities nationwide from being effectively policed. At first, the incentive to reduce expenses might be alluring, since compromises in filtration, purification, and delivery procedures could lead to cost savings that are then transferred to customers.

However, outbreaks of waterborne illnesses would inevitably occur, putting pressure on the government to correct such mistakes. People would rightfully demand access to clean, safe drinking water that is protected by the proper measures, even if modest rate increases or taxes were required to pay for these investments.

Now, let us turn our attention to the unfiltered internet connectivity that takes place in homes and devices daily.

While it facilitates access, commerce, entertainment, and communication, it also allows the flow of malware, fraud, identity theft viruses, hacking tools, bots, phishing scams, and various other cyber threats through unchecked digital pipes. This is permitted despite the serious harm that cybercrime causes; in the last year alone, it cost the world economy over $6 trillion.

This cost encompasses intensive recovery efforts such as rebuilding systems, restoring data backups, legal expenditures, customer notifications, and more. Organizations across the spectrum have invested over $120 billion in cybersecurity defenses—spanning hardware, software, services, and employee training—while attempting to counter the escalating threats.

What is lacking from the current public outcry for action to secure data flows is the urgency typically associated with threats to physical resources, such as air or water. There is also a general sense of helplessness and resignation, where threats are seen as inevitable consequences of engaging in the online world. The challenge faced by internet consumers is starkly different from traditional risks. Boiling suspicious data packets, as one might do with tainted water, is not an option.

These factors contribute significantly to this problem, as does the invisible nature of internet infrastructure and abstract concepts surrounding "data." Most significantly, it reflects a flawed paradigm not found in other contemporary societies: consumers are directly responsible for protecting themselves against sophisticated attacks, which are frequently coordinated by organized cybercrime outfits.

With the majority of adults connected to the internet, often using multiple devices to access various networks daily, the notion of self-securing on such a vast scale becomes implausible as a long-term solution. It would require a full-time commitment comparable to providing one's own electricity or running water—a task far beyond the capabilities of the average user. Yet, despite this, internet service providers (ISPs) maintain a predominantly hands-off approach, granting unfiltered access by default and leaving customers to navigate the treacherous landscape of criminal data flows passing through their infrastructure.

Moreover, the ubiquitous nature of the internet means that younger generations, including children, are increasingly dependent on it for various activities, ranging from online learning to social interactions. However, their lack of proper education about online threats leaves them vulnerable to significant risks. As we grapple with the challenges of securing the internet, it is imperative to develop comprehensive, effective security measures that consider the vulnerabilities faced by users of all ages.

The stark contrast between the rise in cyber threats and the complexities of self-securing underscores the urgent need for change in the digital landscape. The "Way of Water" strategy, which emphasizes control, architecture, and intake valves, emerges as a promising solution to address these concerns.

Many experts today argue that the current state of internet service delivery parallels that of an unregulated utility situation, similar to the polluted rivers of the past before the implementation of environmental regulations. They advocate for stricter control of traffic flow that leverages existing technologies to limit known malicious sources.

Although some contend that this kind of data filtering would violate the right to free speech, security architects and experts do not think much of this claim. Supporters of unrestricted connectivity say that blocking specific IP addresses or traffic might make censorship more effective in suppressing free speech. However, we know with precision which IP addresses currently host the majority of scams, attacks, botnets, malware payloads, and other malicious content. Internet service providers could successfully block these known malicious actors without preventing users from accessing other trustworthy websites or services.

Most experts agree that, similar to how violent criminals lose many of their liberties, anyone who chooses to use internet connectivity as a weapon to steal, defraud, or assault others should not be allowed access to networks that enable them to do harm. Since targeted restrictions on verified threat sources are far less intrusive than many of the current laws governing the sharing of lawful content, cybercrime should not be tolerated under the pretext of "free speech."

Naturally, some technically proficient observers argue that it is already too late to reinvent fundamental elements of the global internet's operation. To these people, overhauling the current infrastructure seems impractical due to sunk costs and the forces of inertia. When one looks at history, however, such arguments quickly become stale.

As the public became more aware of the health risks associated with contamination, we renovated the deteriorating water infrastructure. Therefore, if there is enough momentum to overcome the current economic barriers, society can also redesign some of the fundamental "plumbing" of the internet to address the problem of cyber insecurity.

Mandating that ISPs filter against malware, build robust security into network operations infrastructure, cut off data from confirmed nefarious domains, and address vulnerabilities proactively could drastically strengthen baseline protections for all. Telecommunications trade groups like the Internet Society (ISOC) have outlined frameworks similar to the finance industry's regulations that could responsibly balance security, privacy, and innovation interests when implemented in phases.

The Biden administration recently signaled support for measures along these lines as well, even broaching concepts like a "public option" government-run ISP focused solely on providing the most secure connectivity foundations as a benchmark. Nonetheless, vigorous debate continues around precisely what level of regulation may prove necessary and sufficient absent market pressures. Powerful ISP lobbying groups staunchly resist government control, while activists counter that self-governance has failed for too long already. Both technology and public policy solutions must converge to address such a complex challenge.

Make no mistake—in practice, the main obstacles are still, regrettably, more economic than technological. There is little financial motivation for businesses that profit from open and seamless access paradigms, as they are currently implemented, to change the systems that allow cyber threats to continue.

Again, this brings to mind earlier periods of industrial growth, before environmental or health regulations imposed pollution limits for the benefit of the public, despite profit-driven incentives to dispose of waste carelessly. It is therefore possible to correct the trajectory by using the templates found in policies such as utility regulations. When privately owned infrastructure blatantly disregards its responsibility for the health and safety of its constituents, careful government oversight has demonstrated the ability to turn it around to once again protect these interests.

So what would an actual cybersecurity strategy that made use of the same frameworks as basic utilities look like? First and foremost, we need to acknowledge that significant inbound internet connection infrastructure already exists along America's east and west coasts, enabling global linkage. Just as airlines created the "no-fly list" to prevent high-risk travelers from boarding planes without restriction, internet service providers only need to reinforce and activate firewalls at these critical entry points and update lists of known malicious IP addresses for blocking.

It is feasible for a government organization to maintain an up-to-date "no internet list" using open procedures that permit appeals or disputes, much like domain registrars do. More incoming and outgoing traffic may also be watched for indications of hacker activity, exploits, or flows of stolen data, and high-probability threats can be reported to law enforcement for further investigation.

One important difference between this approach and current models is that the former leaves deeper networks and endpoints connecting individuals relatively untouched. The latter is penetrated by threats that must then be detected and neutralized at endpoints through cloud services or personal antivirus suites. This is frequently after significant damage has already occurred.

Once blacklisted under a "Way of Water" strategy, malicious actors would also find it more difficult to keep switching networks through new channels as the supply of public IP addresses becomes less available worldwide. Although it would not completely eliminate cybersecurity issues, a strategy centered on controlling architecture and intake valves could dramatically lower infection risks throughout the whole internet ecosystem.

But just like people continue to install home filtration systems, buy bottled supplies, and filter water at sinks despite utility cleaning and regulation efforts, so too will reasonable end-user security precautions always be required as long as threats continue to exist and change over time. Risks associated with complex technologies cannot be completely eliminated by a single solution. Nonetheless, to stop the disproportionate burden of those who are least prepared to handle a spreading crisis, it is necessary to demand significant changes in the way the internet's fundamental "pipes" and conduits operate.

Decades ago, we were also ignorant of the health risks posed by widespread pollution and contamination that endangered public water systems. However, after research and experience made those inherent risks clear, society made significant investments to protect communities by putting governance and architectural upgrades into place. This was frequently over the protests of profit-driven corporations that resisted the change.

Today, it is evident that cyber threats regularly cause harm to the real world by taking advantage of the basic tenets of internet architecture. As citizens, it is still our responsibility to push for changes in business practices and engineering standards, as well as to take action on security and safety-related issues.

Long-term integration of ideas like trustworthy and ethical design principles is necessary for the business and engineering processes pertaining to digital systems. For far too long, technology companies have been simply pushing forward with ambitious plans, believing that security could be added later. This approach has clearly failed entire societies. Rather than trying to regulate uncontrollable systems that are entirely accessible to the public, we need to think of consumer-focused tools that have safeguards built in from their very conception.

No technological advance ever remains static for decades without revisions and changes, and at this point, neither should the regulation of the internet. Just as environmental regulations and innovations in water treatment have changed entire industries and led to life-saving public health discoveries, joint efforts to design the internet infrastructure with security in mind could result in significant safety improvements.

The path forward is to combine the fervor of activists with the determination of lawmakers, the know-how of engineers and architects, and the perspectives of those most impacted by insufficient security models. We take the first step toward securing a brighter online future by acknowledging our collective reliance on currently insecure digital systems that provide a necessary human service in the modern era.

Rather than a selection of companies dominating the landscape at the expense of many people's rights and safety, well-thought-out architecture and proactive regulation can provide defense against data waters that have been allowed to run wild for too long. However, it is concerned citizens and consumers who are willing to speak up that are the driving force behind this change, not profit-driven businesses or generally apathetic government bureaucracies.

The vision laid out in this chapter is ambitious. But the severity of daily cyberattacks requires ambitious rethinking of the status quo. We would not allow another nation to openly attack America's infrastructure with impunity. Rethinking internet governance and architecture is imperative if we want the miracle of global connectivity to enable human progress, not undermine it.

Admittedly, there are still complex technical and ethical questions to address regarding this model. How would legitimate domains that are incorrectly blocked be efficiently unblocked? What level of authority should ISPs have to monitor and restrict traffic? How can the appeals process be made fair and transparent?

These are solvable challenges, analogous in many ways to past debates around ethical application of new inventions like automobiles or airplanes. With care and wisdom, solutions can be crafted that reflect our highest shared values.

The Way of Water offers one potential path forward to a more trusted, more secure internet, built on lessons hard won from decades of exponential technological growth outpacing security. The vision sets a direction rather than prescribing every detail. But by raising expectations for internet infrastructure to match its vital place in society, progress becomes possible. The status quo has proven dangerously inadequate as cyber threats multiply. The time for rethinking internet architecture is now upon us, before the metaphorical dams irreversibly collapse.

Chapter 5

PROACTIVE VS. REACTIVE

In the previous chapter, we delved into ambitious ideas aimed at reshaping the very foundations of internet infrastructure and governance in order to combat ever-increasing cyber threats. However, it is important to acknowledge that transformative changes often face obstacles in the form of political resistance and slow adoption. While discussions on more radical strategies continue, organizations are in urgent need of practical approaches to strengthen their current cybersecurity measures.

In recent years, there has been a significant increase in cyberattacks that have caused extensive damage to unprepared organizations. Skilled adversaries have taken advantage of vulnerabilities, resulting in global ransomware damages projected to reach a staggering $265 billion by 2031. These breaches not only come with high costs for remediation but also have a detrimental impact on a company's reputation and the trust of its customers.

Given this growing threat landscape, organizations must make a crucial decision regarding their approach to cybersecurity: should they adopt a proactive or reactive stance?

A proactive cybersecurity program is centered around reducing risks and building resilience, which means anticipating dangers and implementing preventive measures before any attacks take place. A key component is the establishment of a governance, risk and compliance (GRC) framework that is tailored to the organization's specific risks and regulations. This framework enables the development of the policies, controls, resource allocation, and role assignments necessary to effectively execute a cybersecurity strategy.

Conducting comprehensive risk assessments is vital in a proactive approach. By identifying valuable assets, weak points and risk tolerance, a company can prioritize resources and put appropriate security measures into place. Categorizing data and assets based on their sensitivity allows organizations to provide tailored safeguards for critical information. Vendor risk programs ensure that partners and suppliers adhere to security standards.

Proactive cybersecurity also involves continuous vulnerability monitoring, strengthening defenses preemptively, and preparing for incident response. This includes setting up multilayered controls such as next-generation firewalls, virtual private networks (VPNs), endpoint detection, and data encryption.

Leveraging advanced threat detection technologies, such as AI, blockchain, and quantum sensing, helps identify anomalies early and minimize potential damage. Some organizations even employ ethical hackers to proactively identify weaknesses in their networks. Incident response planning focuses on minimizing damage and recovery when preventive measures fail.

On the other hand, reactive cybersecurity involves responding to incidents after they have occurred. While reactive responses can lead to improvements in defenses, they tend to be more costly and disruptive. When threat detection is delayed, adversaries can establish a stronger presence, making it harder to contain and minimize potential harm.

To be a leader in cybersecurity, it is important to have a combination of proactive defenses and reactive incident response capabilities. Proactive measures help reduce the frequency of attacks and vulnerabilities, while reactive strategies help minimize the impact of incidents. By having a dynamic proactive program supported by effective reactive capacities, organizations can build resilience in the face of evolving dangers.

Just like strong buildings require solid foundations, effective security starts with well-defined policies. Cybersecurity policies serve as the backbone of an organization's security posture, acting as a living document that evolves with the changing threat landscape. They establish standards, controls, and procedures, providing a framework for the executive oversight, resource allocation, and role assignments necessary to implement a robust cybersecurity strategy.

Whether for large enterprises or small businesses, it is crucial to have comprehensive policies tailored to specific risks. Key elements of these policies include conducting thorough risk assessments to identify critical assets and vulnerabilities, classifying data and assets based on their sensitivity, implementing role-based access controls to limit unauthorized access, and outlining clear incident response procedures.

Regular cybersecurity audits and risk assessments are necessary for proactive vulnerability identification and mitigation. Audits assess system configurations, patch status, and response, while risk assessments utilize qualitative and quantitative methods to analyze threats, their likelihoods, and the potential impacts on the business. Continuous monitoring and updating of these audits and risk assessments is essential as new dangers emerge.

To protect against external threats, organizations must have robust network security measures in place. Firewalls, intrusion detection and prevention systems (IDS/IPS), and network access controls filter traffic and detect and halt potential hazards. Segmenting networks into distinct tiers restricts access between them, effectively containing breaches and limiting the scope of a cyberattack.

Another crucial component is secure remote access for employees, especially in the era of remote work. Virtual private networks (VPNs) and multifactor authentication are important for secure remote access, as they help prevent unauthorized entries and data breaches. As with other protections, it is necessary to regularly update and upgrade these controls to maintain strong network defenses against evolving threats.

Endpoint protection is also vital because endpoints like laptops, mobile phones, and POS terminals are often the primary targets in cyberattacks. Antivirus software, anti-malware tools, and prompt updates help defend against known dangers. Endpoint detection and response (EDR) solutions use behavioral analytics to identify advanced threats that traditional methods may not easily spot.

By implementing security measures for every endpoint in an organization, including company-owned devices and bring-your-own-device (BYOD) platforms, the attack surface can be significantly reduced. This comprehensive approach enhances overall cybersecurity resilience and protects against a wide range of potential cyberattacks.

The above network protections and security practices work together to create strong defenses against cyber threats. In this way, organizations can improve their security and reduce the success rate of attacks, while continuously monitoring and updating security measures ensures ongoing protection against evolving adversaries.

As we work towards establishing a security-focused organizational culture, we must also recognize the need to educate and engage the younger generation in cybersecurity awareness. With technology increasingly integrated into the lives of children and young students, fostering a strong foundation in online safety, digital citizenship, and responsible internet use is essential.

Schools should incorporate cybersecurity education into their curricula, covering topics such as safe online behavior, strong password practices, and threat identification. By empowering young learners with this knowledge, we can help them to navigate the digital landscape.

Parents also play a vital role in shaping their children's online experiences. Encouraging open dialogue, setting clear boundaries, and utilizing parental control tools can help create a safe digital environment at home. Additionally, parents should be informed about the possible risks associated with child-themed devices, such as data collection practices and security concerns.

By involving the younger generation in cybersecurity awareness initiatives, we can build a culture of cyber safety and responsibility. This proactive strategy will ensure that future generations are prepared to traverse our increasingly digital world with confidence and vigilance, ultimately contributing to a more secure online environment for all.

In order to establish a security-focused organizational culture, a comprehensive approach is needed, which involves engaging employees and promoting shared responsibility for cybersecurity. One important aspect is incorporating security training and awareness into the onboarding process for new employees, so that security is prioritized from the very start of their journey with the organization. Regular refreshers should also be provided to all staff to reinforce security habits and keep them informed about the latest threats and best practices.

Interactive training modules, games, and attack simulations help employees develop a better understanding of threat identification, policy adherence, and password hygiene. This makes security practices more intuitive and actionable. Visual aids such as posters placed in common areas serve as reminders, while brown bag talks encourage discussion and deeper understanding of relevant topics.

It is important to promote a culture where everyone takes responsibility for security, not just the IT department. Sharing security information and reminders in common areas like break rooms makes employees feel valued and involved in their employer's cyber resilience. Creating an open and supportive environment encourages staff to voice concerns without fear of blame, allowing organizations to benefit from the collective knowledge and vigilance of their workforce.

Unintentional insider threats, such as misconfigured firewalls, unsecured laptops, or weak password practices, often occur due to human error. Implementing change control oversight helps prevent misconfigurations, while limiting access privileges ensures that individuals only have the level of access they need. Additional measures like mandatory VPN connections provide extra security for remote system entry.

To detect and manage rogue insiders, organizations can apply user behavior analytics to identify unusual access attempts or behaviors that may indicate malicious intent. Privileged access management tools can control and monitor high-level accounts, while separation of duties means that no single user has unchecked ingress to critical systems. Regular access reviews help promptly identify unauthorized entry, and swift off-boarding processes safely revoke former employees' access rights.

By cultivating a security-focused culture and practices, organizations strengthen their defenses against both unintentional and intentional insider threats. Employees will become more aware, engaged, and proactive in maintaining a secure environment, ultimately contributing to the overall resilience of their employer's cybersecurity efforts.

In addition, it's imperative to have strong protections for data at rest and in transit to keep sensitive files, databases, emails, and virtual machine images safe from unauthorized access or changes. These protections are the foundation for preserving valuable data.

Regularly backing up important data is essential to be able to restore it in case of malware attacks or system failures. Backup copies should be stored offline to avoid the same risks as online systems. It is also important to periodically test the recovery process to ensure that backups are usable and effective.

The 3-2-1 rule is a good guideline for data backup and storage. This rule suggests maintaining at least three copies of data, storing them on at least two different types of media, and keeping at least one copy offline. Creating immutable backups can also help prevent encryption attacks on backup files by making them unalterable and resistant to unauthorized changes.

In order to avoid hefty fines and safeguard the company's reputation, data protection laws and compliance requirements must be met. Regulations like GDPR, NYDFS, HIPAA, and CCPA impose strict rules for data protection. To meet these obligations, organizations should implement measures such as data classification, access controls, encryption, audit logs, consent management systems, and breach notification procedures. Non-compliance not only results in financial penalties but also erodes consumer and partner trust.

While organizations should proactively strive to prevent breaches, they should also be prepared for them as a possibility. Establishing an incident response (IR) team, providing them with training, and defining communication protocols are crucial steps in preparing for security incidents. Creating notification templates in advance helps streamline communication during a crisis. Conducting simulated breach scenarios allows for testing the effectiveness of incident response capabilities, and analyzing failures through postmortems helps identify areas for improvement.

Thoroughly investigating the root causes of attacks helps organizations identify any missed alerts, understand how adversaries accessed their systems, and uncover any weaknesses that allowed the attack to progress. Remediation actions, such as technology upgrades, policy changes, and additional staff training, can then be taken to address vulnerabilities and enhance security.

Through the implementation of continuous data protection best practices and robust incident response capabilities, companies can effectively identify risks, appropriately respond to incidents, and quickly recover from cyberattacks while minimizing the associated damage.

To enhance cybersecurity measures and proactively address emerging dangers, organizations must harness the power of cutting-edge technologies. For instance, AI-based behavior analytics solutions can scrutinize patterns in user behavior to detect anomalies that may indicate insider threats or account takeover attacks. By establishing baselines and identifying deviations, these solutions help to flag possible malicious activities.

Similarly, advanced malware prevention and detection algorithms surpass traditional antivirus offerings by recognizing complex code signatures. Forward-thinking methods like these enable companies to identify and mitigate evolving threats before they can cause considerable harm.

When integrating AI and analytics solutions, organizations must also prioritize privacy concerns. Clear communication is essential to highlight the benefits of these technologies and maintain transparency regarding the data collection necessary for training algorithms. This strategy fosters trust and ensures compliance with privacy regulations.

In the realm of cybersecurity, blockchain technology has emerged as a game-changing innovation for bolstering data security and thwarting manipulation attacks. By creating an immutable, distributed ledger of transactions, blockchain guarantees that records cannot be tampered with, thus introducing a layer of trust and transparency to digital systems.

As you'll discover in greater detail later in this book, blockchain operates through a decentralized network of nodes or computers that validate and record data in blocks. These blocks are then linked together using cryptography, making it nearly impossible to alter or delete the information stored within them.

The adoption of blockchain technology allows organizations to secure their data and systems, particularly in areas like supply chain management, voting systems, and DNS infrastructure, where data integrity and authenticity are paramount. This groundbreaking approach holds the potential to transform the cybersecurity landscape and tackle some of the most pressing challenges in the digital age.

However, there are challenges to overcome for widespread blockchain adoption, including scalability, complexity, and sustainability. Addressing vulnerabilities in the underlying encryption schemes is likewise crucial to ensure the security of blockchain implementations. But by tackling these challenges head-on, organizations can unlock the full potential of blockchain technology in strengthening their cybersecurity posture and safeguarding against emerging threats.

Quantum computing is another advanced technology that can enhance cybersecurity measures. However, it also brings both risks and opportunities. Current cryptographic schemes used in blockchain, TLS certificates, and identity management systems may become vulnerable to quantum computing methods. To protect against its possible algorithm-breaking capabilities, organizations should transition to quantum-resistant cryptography.

On the other hand, quantum sensing technology has the potential to monitor critical infrastructure and data centers for signs of cyber intrusions. By detecting tiny disturbances in quantum states, alerts can be triggered if there are attempts to physically tamper with hardware. Early detection of such tampering allows organizations to prevent extensive breaches.

As quantum computing capabilities continue to develop in the next decade, post-quantum cryptography and communication will be crucial for resilience. Quantum key distribution offers perfectly secure key exchanges compared to traditional public key infrastructure. Quantum-safe encryption also provides stronger encryption algorithms that are resistant to decryption attempts from quantum computers.

These innovative technologies enable organizations to detect threats faster, respond earlier, and establish hardier defenses before attacks escalate. Through the use of AI, blockchain, and quantum technology, organizations can gain a significant edge over their digital adversaries.

In the rapidly evolving digital landscape, continuous cybersecurity research and development are vital for understanding and addressing new threats. As technology advances, so must our strategies for protecting digital systems and sensitive data. To build robust defenses, we must develop dynamically adaptive policies, maintain vigilant staff, establish supportive infrastructure, and adopt cutting-edge technologies capable of withstanding future challenges.

Commitment to ongoing education and training is a key component of a strong cybersecurity strategy. Cybersecurity professionals, as well as users of digital systems, must stay informed about the latest threats, technologies, and best practices. By fostering a culture of continued learning and improvement, we empower individuals and organizations to proactively address new risks and maintain a strong defensive posture.

Investment in research, development, and education is likewise paramount for creating a secure and resilient digital environment. The constant evolution of cybersecurity strategies is not just a choice but a necessity in today's interconnected world. Through our collective commitment to innovation and learning, we can stay ahead of potential threats and work together to protect the digital landscape for generations to come.

As we conclude this chapter, it is essential to remember that the fight against cyber threats is an ongoing battle. Cybercriminals continuously evolve their tactics, finding new ways to compromise even the most robust security systems, and organizations must adapt their defenses to stay ahead. The consequences of failing to do so include a cascade of damaging consequences, extending far beyond the immediate financial losses and disruptions to business operations.

One of the most significant impacts of a cyber breach is the erosion of trust among customers, partners, and stakeholders. This loss of confidence can result in substantial reputational damage, undermining an organization's brand and potentially leading to a decline in market share, revenue, and shareholder value.

Moreover, the Chief Information Security Officer (CISO) in charge at the time of the incident may face severe consequences, including disciplinary action, dismissal, or even legal repercussions. The ramifications can be career-ending, further emphasizing the critical importance of effective cybersecurity leadership.

Organizations must prioritize cybersecurity at all levels, from the boardroom to the individual employee. Robust defenses, regular risk assessments, and a culture of continuous learning are vital to a comprehensive cybersecurity strategy. By investing in these measures, organizations can better protect themselves, their customers, and the broader digital economy from the damaging effects of cyber breaches.

In the context of the "Cyber Race" and "Cybersecurity Endgame," the concepts discussed in this chapter highlight the importance of adopting a proactive security stance. As organizations face mounting pressure to protect against increasingly sophisticated cyber threats, they must recognize that a reactive approach may leave them vulnerable and constantly on the back foot.

The Cyber Race is a continuous competition between organizations and cybercriminals, with each side striving to outmaneuver the other. By adopting a proactive approach, organizations can better position themselves in this race, anticipating potential threats and addressing vulnerabilities before they can be exploited. This not only helps to minimize the risk of successful attacks but also enables organizations to stay ahead of the curve in an ever-evolving threat landscape.

Furthermore, a proactive approach aligns with the overarching goals of the Cybersecurity Endgame, which is to create a secure and resilient digital environment. By focusing on prevention, organizations can contribute to strengthening the broader cybersecurity ecosystem. This approach recognizes that the actions of individual organizations affect the security landscape as a whole and emphasizes the importance of collaboration and shared responsibility in achieving the desired end.

While reactive measures will always have a place in an organization's cybersecurity strategy, it is essential to prioritize proactive efforts in order to gain an advantage in the Cyber Race and work towards the ultimate Cybersecurity Endgame. By embracing a proactive stance, organizations can better protect their assets, maintain trust with customers, and contribute to the development of a more secure digital world.

As we move forward, we must remain vigilant and committed to learning about the latest threats and best practices in cybersecurity. By doing so, we can collectively create a safer, more secure digital landscape for all.

In the next chapter, we will delve into the topic of ransomware, a particularly severe menace in the world of cybersecurity. As organizations face such growing risks, it is crucial to understand the specific strategies, tools, and techniques that can help bolster their cybersecurity posture against these malicious attacks.

Chapter 6
RANSOMWARE

In a medium-sized accounting firm in Ohio, the unexpected shutdown of computer systems initially seemed like a random technical glitch. However, it quickly escalated into a crisis when employees found themselves unable to access their files and encryption alerts appeared on their screens.

It soon became evident that the firm had fallen prey to the Conti ransomware gang. These cybercriminals demanded a hefty sum of $250,000 in Bitcoin to release the hijacked systems and data. With crucial tax reporting deadlines looming, the firm had no choice but to give in to the ransom. Consequently, sensitive client financial records and valuable intellectual property fell into the hands of cybercriminals. Sadly, this incident is just one example of the alarming surge in ransomware attacks worldwide.

As a certified Incident Commander with the Department of Homeland Security, I have witnessed firsthand the devastation these attacks inflict on organizations of all sizes. From global corporations to small businesses, no organization is immune to the threat of ransomware. In several cases, I have been called upon to lead recovery efforts after clients were hit by ransomware campaigns, regardless of their size or industry. My role involves liaising with the FBI, enacting containment protocols, and guiding clients through the necessary steps to restore operations with minimal brand damage. While the specific organizations must remain confidential, this frontline experience has shown me how disruptive attacks can become if companies are not prepared.

Ransomware attacks, which exploit encryption to extort victims, have skyrocketed by over 500% since 2018. They have transformed into a highly lucrative criminal enterprise, generating billions of dollars for cyber gangs and wreaking havoc on businesses, governments, hospitals, and everyday internet users.

Initially emerging in the mid-2000s, ransomware was a novel method for extorting money from victims in the underground world of cybercrime. Over time, these attacks have become even more devastating, causing widespread paralysis of systems and infrastructure on a global scale. Ransomware has now evolved from isolated incidents into a significant and urgent cyber threat, intertwined with modern geopolitical tensions and conflicts between nations.

Highly sophisticated hacking groups like Conti, REvil, DarkSide, and others execute high-impact attacks that cripple companies. Additionally, the availability of ransomware-as-a-service (RaaS) offerings has made it easier for inexperienced hackers to launch large-scale infection campaigns. As a result, victims are faced with difficult choices: pay the ransom or endure disruption and permanent loss of data.

Even nation-state actors have started outsourcing cyber operations to ransomware groups. Their goal is to economically and socially destabilize adversary nations through digital means. The use of cryptocurrencies for anonymous ransom payments further complicates global law enforcement efforts to prevent attacks.

Cryptocurrencies like Bitcoin and Monero have become closely associated with most ransomware schemes. Their pseudo-anonymous nature allows adversaries to receive untraceable payments. Dark web marketplaces also aid in laundering ransom money and hiding transactions from authorities.

These developments have made ransomware a top concern in the Cyber Race, as governments, businesses, and infrastructure operators struggle with questions about how to manage the unrelenting scourge of extortionist hackers. Unfortunately, as of today there are still no easy solutions to this problem.

This chapter delves into the ransomware landscape, exploring the main individuals involved, troubling developments such as nation-state support, the impact on the interconnected global economy, and strategies for combating this urgent threat. By understanding these factors and their interplay, organizations can better position themselves in the Cyber Race and contribute to the broader goal of achieving a secure and resilient digital environment in the Cybersecurity Endgame.

Ransomware attacks have become a lucrative business for cybercriminals, with the potential for high financial rewards and minimal risk. Hackers employ a two-stage process designed to exert maximum pressure on victims and increase the likelihood of payment.

During the initial stage, adversaries employ various tactics to breach the systems of companies or agencies. Phishing emails, which trick unsuspecting users into clicking on malicious links or downloading infected attachments, are a common method used to gain initial access. Exploiting vulnerabilities in remote desktop protocol (RDP) or unpatched software is another avenue. These vulnerabilities provide an entry point for hackers to infiltrate the target's network.

Once inside the system, the hackers gradually escalate their privileges, moving from initial access to complete control. This process allows them to navigate through the network, identify valuable data and locate backup repositories. By encrypting files and compromising backup systems, they ensure that the victim's data is effectively held hostage.

With the data encrypted and under their control, the criminals demand a ransom from the victim. The requested amount is often substantial and is typically demanded in cryptocurrencies such as Bitcoin, which are difficult to trace. Recent trends indicate a significant increase in the average ransom payment, showing the growing profitability of these attacks.

If the victim refuses or is unable to pay the ransom, the adversaries resort to additional tactics to increase pressure. They may threaten to publicly expose the stolen data, potentially damaging the victim's reputation or causing legal and regulatory consequences. Alternatively, they may choose to sell the data on dark web markets, further exploiting its value.

Faced with these threats and the potential disruption to their operations, many victims reluctantly choose to pay the ransom. This decision is often made without any guarantee that their attackers will honor their end of the bargain and provide the decryption key.

The impact of ransomware attacks is expected to continue growing in the coming years. Cybersecurity Ventures predicts that global damages from ransomware could exceed $265 billion annually by 2031. This projection reflects the increasing number of attacks and the growing sophistication of the techniques used.

One factor contributing to the rise in attacks is the emergence of ransomware-as-a-service (RaaS). RaaS involves experienced hackers allowing inexperienced threat actors to participate in ransomware schemes by providing them with access to tools, infrastructure, and support. This decentralized approach has lowered the barrier to entry, enabling financially motivated individuals with limited technical expertise to engage in digital extortion.

The continuous advancement of ransomware techniques and the lucrative financial gains associated with it make it unlikely that criminals will lose their motivation in the near future. Unless significant changes occur in the global digital landscape, this trend is expected to persist.

While most ransomware campaigns are driven by profit-seeking cyber gangs, there are concerning indications of nation-state actors collaborating with these groups to target their perceived enemies. Experts have found links between well-known ransomware groups like Conti, REvil, and Ryuk, and Russian intelligence services. The FBI directly attributed the 2017 WannaCry attack, which affected 300,000 computers worldwide, to North Korean state hackers. Conti significantly impacted Ireland's healthcare system during the Russia-Ireland diplomatic tensions related to Ukraine. Additionally, the DarkSide group, which mysteriously vanished after the Colonial Pipeline attack, showed signs of being a state-proxy entity.

These instances highlight the concerning trend of nation-states collaborating with cybercriminals who share their national interests. This approach allows for significant disruption capabilities with an asymmetrical advantage, while also enabling the perpetrators to distance themselves from the ransomware activities. Covert alliances like this make it extremely challenging to implement effective global mitigation strategies.

But this blurring of lines between financially motivated cybercrime and government-sponsored operations violates norms for responsible peacetime state actions in cyberspace. Ransomware groups with proven nation-state connections arguably constitute cyberterrorism threats given their indiscriminate targeting of civil infrastructure.

Countering this dangerous status quo requires multinational diplomatic and law enforcement efforts to establish consequences for sponsoring ransomware groups, even via informal outsourcing. Otherwise, unchecked proxy disruption will continue escalating as weaker nations employ it against stronger foes.

Beyond the millions of dollars lost directly to ransom payments, ripple effects from major attacks frequently send economic shockwaves across interconnected sectors and geographies. These second-order impacts often far exceed the initial ransom amounts.

Cybersecurity Ventures predicts that ransomware could cost the global economy over $30 billion per year as cascading business interruption and recovery costs accumulate after large attacks. Such estimates remain difficult to accurately quantify, but the immense financial damages are apparent across impacted countries.

In the United States, healthcare, manufacturing, food production and other sectors suffer steep annual losses, especially when ransomware rapidly spreads through supply chain networks. Developed nations in Europe have also encountered disruptions in healthcare and critical infrastructure, resulting in an estimated combined damage of over $10 billion throughout the continent.

Asian critical infrastructure and businesses have not been spared from attacks either, despite a higher level of public awareness regarding the global emergence of ransomware as a severe threat.

Developing countries often suffer even worse proportional harm, with ransomware undermining advancements in IT infrastructure and digital services. India is still struggling with fallout from the 2017 WannaCry attack that affected over 100,000 computers. South American hospitals were debilitated amidst the COVID-19 crisis by cyber gangs showing no restraint.

While insurance coverage can offset some financial impacts, premium hikes after major incidents counteract this benefit. Ransomware also poses grave risks for financial system stability if banking networks are disrupted. Realistic economic evaluations must consider these spiraling second-order effects.

Given ransomware's complex cross-border dimensions, reducing the mounting threat requires coordinated action across both public and private sectors, as well as internationally. Cyber resilience practices, modernized IT systems, adversary attribution, and supply chain security must all be collectively prioritized.

Many experts advise organizations against paying ransoms, although realistically this option remains on the table for companies lacking backups or facing bankruptcy. Cyber insurance can offset costs, but should not be the primary mitigation strategy.

All organizations, no matter the size, should proactively implement offline backups, multi-factor authentication, and end-to-end encryption, as well as technologies like firewalls, antimalware programs, and endpoint detection and response solutions to harden environments against intrusions. Regular vulnerability assessments and penetration testing should also be done to identify security gaps.

Additionally, ongoing employee education and attack simulation drills build organizational vigilance and preparedness. Clear incident response plans ensure teams can act swiftly in the unfortunate event of a real ransomware incursion.

Collaboration is also key in the fight against ransomware. Organizations must share threat intelligence, best practices, and incident response strategies to strengthen collective defenses. Public-private partnerships between government agencies, industry experts, and cybersecurity firms can facilitate the exchange of information and resources, enabling swift and coordinated responses to each new danger.

Looking forward, the ultimate solution to ransomware lies in dismantling the underlying extortion model's profitability. By implementing a combination of deterrent measures, financial system controls, cryptocurrency regulation, and international collaboration, we can eventually confine this hazard to the periphery from which it emerged.

However, until this ideal world becomes a reality, ransomware continues to pose a significant and immediate threat to businesses, governments and individuals worldwide. Ransomware as a consequence of cyber breaches demands a comprehensive and unwavering commitment to cybersecurity. Organizations must therefore recognize that prevention alone is not enough; they must also invest in robust incident response plans, data backup strategies and recovery mechanisms.

Chapter 7

SECURITY INCIDENT VS. SECURITY BREACH

In our exploration of the complex world of cybersecurity, we've discussed various threats, including the damaging impact of ransomware. While ransomware poses a significant danger, it's not the only type of cyberattack organizations face. Cyberattacks come in various forms, each with different objectives, methods, and impacts. Some aim to steal information, upset systems, extort money, or manipulate behavior. They range from sophisticated and targeted to simple and opportunistic. Certain attacks are noticeable and disruptive, while others are carried out covertly and go undetected.

Among these variations, security teams need to know the distinction between security incidents and security breaches. Although these terms may seem interchangeable, accurately categorizing such events significantly affects the response. Understanding what sets a security incident and a security breach apart is crucial for organizations to effectively manage and mitigate cybersecurity risks. In this chapter, we'll define these two terms, provide examples, and explain the importance of recognizing their differences.

A security incident is any event that compromises the confidentiality, integrity, or availability of an information system. Unauthorized login attempts, malware infections, or even ping floods are all considered security incidents. However, the occurrence of an attack doesn't necessarily mean that data was stolen. The key point is that there was a threat to the system, but no confirmation of theft or damage.

On the other hand, a security breach is more serious, carrying legal, regulatory, reputational, and financial consequences. It occurs when sensitive information or assets are exposed, stolen, changed, or deleted without permission. For example, if a hacker uses a Structured Query Language (SQL) injection attack to steal customer credit card data, it's considered a breach. Similarly, if an insider uploads company financials to a cloud storage service like Dropbox without authorization, that also qualifies as a breach. Data exfiltration, manipulation, or destruction has occurred in both cases.

To simplify, a security incident threatens the system but doesn't result in data loss, while a security breach involves compromised data. To illustrate this difference, let's examine a well-known security incident that occurred a few years ago.

The Equifax data breach, which occurred in 2017, is a high-profile case that highlights the severe consequences that can arise when security incidents escalate into full-blown breaches. Equifax, a consumer credit reporting agency, revealed that the personal information of approximately 143 million consumers had been compromised. This information included personal details such as names, Social Security numbers, birth dates, addresses, and, in some instances, driver's license numbers.

The incident began when adversaries exploited a vulnerability in the Apache Struts web application framework, which Equifax was using in one of its web applications. This vulnerability had been disclosed and patched months before the breach occurred, but Equifax had failed to apply the necessary updates. As a result, the adversaries gained unauthorized access to Equifax's systems and were able to exfiltrate consumer data.

This example displays the importance of differentiating between security incidents and security breaches. While the initial unauthorized access and exploitation of the vulnerability can be considered a security incident, the exfiltration of sensitive data elevated the situation to a security breach. The breach resulted in significant financial and reputational consequences for Equifax, as well as identity theft and fraud risks for affected consumers.

Organizations must learn from examples like this and ensure they can accurately classify and respond to cybersecurity threats. Recognizing the differences between security incidents and security breaches enables them to implement appropriate response measures and effectively utilize resources, thereby minimizing the impact on their systems and data.

The consequences of misclassifying security events can be severe, leading to inadequate responses, increased damage, and potential legal and regulatory repercussions. If an organization incorrectly identifies a security breach as a mere security incident, it may not allocate sufficient resources or take necessary steps to mitigate the impact, investigate the root cause, and prevent future occurrences.

This oversight could leave the organization vulnerable to further attacks and compromise, as adversaries may still have access to the network or may have installed additional malware or backdoors. Furthermore, the organization may fail to notify affected individuals or regulatory authorities, resulting in legal penalties and reputational damage.

Conversely, if an organization incorrectly classifies a security incident as a breach, it may cause unnecessary panic, the allocation of excessive resources, or negative publicity. This could divert valuable resources from other critical security tasks and potentially harm the company's reputation if it prematurely discloses a presumed breach that did not occur.

A security breach can have devastating consequences for an organization and its stakeholders, as demonstrated by the Equifax case. Therefore, security teams must catch security incidents as early as possible to moderate potential damage. Rapid detection allows organizations to contain, analyze, and respond to threats more effectively. It can also help reduce the impact of an event and prevent it from escalating or spreading.

However, detecting security incidents and breaches is a challenging task. Security teams face numerous obstacles, such as collecting and analyzing data from various sources, addressing different types of threats, and managing resource limitations. The increasing volume and complexity of security incidents and breaches also presents difficulties, exacerbated by the high demand for qualified security professionals and the prevailing skills gap in the workforce.

Security teams rely on various tools and technologies, including firewalls, antivirus software, intrusion detection systems, security information and event management systems, and threat intelligence platforms, to help spot security incidents and breaches. However, these tools can have limitations, such as false positives, false negatives, compatibility issues, configuration errors, and outdated information.

To overcome these challenges and improve detection capabilities, security teams must adopt a proactive, comprehensive approach to security monitoring and analysis. This begins with establishing and prioritizing security goals, requirements, and measurements. Organizations should align their security objectives with their business goals and prioritize the protection of critical assets, data, and systems.

By implementing robust security policies, procedures, and controls, organizations can enhance their ability to diagnose, respond to, and recover from security incidents and breaches. Regularly reviewing and updating these measures ensures their continued effectiveness in the face of evolving threats and technological advancements. Furthermore, fostering a culture of security awareness and ongoing improvement within the organization can help employees recognize and report potential issues, contributing to a more resilient security posture.

Organizations must use a structured approach that includes several key steps to manage and respond to security incidents and breaches. One of these steps is the implementation and enforcement of security policies, procedures, and standards. Establishing clear guidelines that dictate expected security practices and behaviors is vital, as is communicating them to staff, partners, and customers and enforcing them through regular audits, reviews, and penalties.

Another critical step is the deployment and integration of security tools and technologies. Organizations must select and implement appropriate options that aid in collecting, processing, analyzing, and visualizing security data. Ensuring that these tools and technologies are compatible, interoperable, scalable, and up-to-date is essential for maintaining a robust security posture. In addition, organizations must consider the entire attack surface and the entire security lifecycle when selecting their security solutions.

The continuous development and maintenance of security skills and knowledge is yet another vital aspect of managing security incidents and breaches. Organizations should prioritize keeping their security teams informed about the latest trends, techniques, and best practices in the cybersecurity landscape. Regular training and education, as well as seeking guidance from external sources such as security experts, communities, and organizations, are imperative for staying ahead of emerging threats.

Collaboration and coordination with both internal and external stakeholders is a further necessary step. A culture of information sharing and resource coordination must be fostered between internal teams and departments, such as IT, legal, compliance, and public relations. Additionally, organizations should collaborate and exchange information with external parties, including vendors, customers, peers, regulators, and law enforcement.

Upon detecting a security incident or breach, organizations must report it to the appropriate authorities. Reporting is a crucial part of security management, as it helps maintain compliance with relevant laws and regulations, informs affected parties, and enables them to seek assistance and support from external sources. But reporting security incidents and breaches can be complex, requiring organizations to consider factors such as timing and frequency, content and format, and audience and communication channels.

Timing and frequency of reports are the first key factors. Organizations must strike a balance between reporting incidents and breaches promptly and ensuring the accuracy and completeness of the information provided. As new information emerges or the situation evolves, they should update and revise their reports accordingly.

Second, organizations should report security incidents and breaches clearly and concisely, while still providing sufficient detail and a comprehensive overview. The information included in the report should cover the nature, scope, impact, and cause of the event, as well as actions taken, planned actions, and any recommendations or lessons learned. Adherence to specific reporting format requirements and standards, such as language, structure, and style, is essential.

Third, organizations must disclose incidents or breaches to the appropriate individuals at the appropriate levels and through the appropriate outlets. Stakeholders who need to be informed may include senior management, the board of directors, legal counsel, regulators, customers, the media, and the public. The choice of communication channel, whether it be email, phone, web, or social media, should be based on availability, reliability, and security considerations.

By following these structured steps and considerations, organizations can improve their ability to effectively manage and respond to security issues, ultimately strengthening their cybersecurity posture and minimizing the potential impact on their operations, reputation, and stakeholders.

Understanding and addressing the legal and ethical implications of reporting security incidents and breaches is a critical consideration in the reporting process. Security teams must ensure their reporting complies with applicable laws and regulations while adhering to the organization's ethical principles and values. This requires awareness of the organization's legal obligations and rights, as well as those of the individuals involved. Key legal aspects include notification, disclosure, and consent requirements, along with privacy, confidentiality, and liability protections. Ethical considerations demand honesty, transparency, and accountability in reporting, avoiding any misleading, inaccurate, or incomplete information.

Navigating these factors and challenges effectively requires security teams to adopt a systematic and strategic approach. This entails developing and implementing a comprehensive security reporting policy and plan that clearly defines the related objectives, scope, roles, and responsibilities. Aligning this policy with the organization's mission, vision, and values and communicating it to all relevant stakeholders leads to a shared understanding and commitment to security reporting best practices.

Establishing and maintaining a dedicated security reporting team and process is another essential aspect of this approach. The team, composed of qualified individuals well-trained in handling security reporting tasks, should follow a process that includes gathering, analyzing, verifying, and disseminating information. Continuously monitoring and evaluating the team and process performance will allow organizations to make improvements and adjustments as needed.

Leveraging appropriate security reporting tools and technologies is also crucial to enhance the collection, processing, analysis, and visualization of security data through dashboards, reports, and alerts. Furthermore, utilizing tools and technologies that bolster the secure communication and sharing of security information, such as encryption, authentication, and authorization measures, is of utmost importance.

Lastly, security teams should actively request feedback and learn from the outcomes of their reporting efforts. By establishing a mechanism for receiving feedback and promoting a culture of continuous learning, organizations can significantly enhance the quality and usefulness of security reporting. Seeking feedback from the organization and stakeholders, as well as learning from the consequences and repercussions of reporting, enables organizations to improve their security posture and resilience over time.

Once a security incident or breach has been reported, the focus shifts to investigating and resolving the issue. This involves identifying and addressing the root cause of the event and restoring the organization's normal operations.

However, several challenges may arise during the investigation and resolution process.

One possible challenge is a lack of information or evidence regarding the event. Security teams may struggle to determine who carried out the attack, their motives, the methods used, and the extent of the damage or data compromise. In some cases, any available information or evidence may be unclear or contradictory, further complicating the investigation process.

Another challenge arises if any of the evidence is changed or disappears. Adversaries, system processes, or environmental factors may alter, delete, or corrupt digital evidence. Similarly, physical evidence may be damaged, lost, or contaminated during the incident response and recovery process. This has even more potential to muddy the proceedings and compromise the accuracy and completeness of the findings.

Investigating and resolving security incidents and breaches becomes particularly complex when dealing with diverse and interconnected systems. Security teams may need to work with various affected systems, such as computers, software, networks, cloud services, mobile devices, and IoT devices. These all may exhibit different characteristics, vulnerabilities, and interactions, making it difficult to pinpoint the root cause and implement effective solutions. As a result, organizations must adopt a holistic approach to incident response, taking into account the interplay between different networks and components.

As we have seen, distinguishing between security incidents and breaches is crucial to accurately assess cyber events and implement appropriate response plans. The distinction helps organizations prioritize their response efforts, allocate resources efficiently, and minimize any damage. Meanwhile, enhancing threat detection capabilities allows organizations to properly respond to and moderate both security incidents and breaches.

Preventing and mitigating the impact of security breaches requires a defense-in-depth approach that combines prevention, detection, response, and recovery strategies. This includes instituting robust access controls, encryption, and data loss prevention (DLP) solutions, as well as conducting regular vulnerability assessments and penetration tests to identify potential weaknesses in the organization's security defenses. This will be explained more in Chapter 14 of this book.

However, it is crucial to recognize that no organization can eliminate all risks, even after implementing strong prevention measures. Every organization remains vulnerable to potential disruptions caused by cyberattacks, technology failures, or natural disasters. This is why it’s so important to have a business continuity and disaster recovery plan in place to ensure the ongoing operation of critical business functions during major disruptions. By proactively establishing continuity arrangements, organizations can maintain essential operations even in the face of unpredictable events. The ability to adapt and respond effectively to these challenges is a hallmark of resilience, requiring a comprehensive and well-practiced incident response strategy.

In the upcoming chapter, we will delve into the concepts of business impact analysis, recovery strategies, and best practices for continuity planning. By exploring these topics, organizations will be better equipped to deal with inevitable disruptions. Much like how medical professionals stress the importance of preventing illnesses while also preparing for disease outbreaks, cybersecurity leaders must be prepared with both strong defenses and robust contingency plans to face an uncertain future.

Chapter 8

BUSINESS CONTINUITY AND DISASTER RECOVERY (BCDR) PLANNING

Imagine we run a small online gift business, selling customized products worldwide. We've invested years building our website, brand, and 50,000 loyal customers. Our business depends on smooth website function, secure servers, and reliable networks.

One morning, we wake up to a nightmare. Hackers have attacked our site, stolen customer payment data, posted offensive images, and disabled orders and shipping. Within hours, thousands of angry calls and emails flood in demanding refunds and threatening lawsuits.

We quickly realize this is no trivial incident. The hackers deleted product databases, corrupted backups, and disappeared without a trace. Specialists estimate it will take weeks and $2 million to fully restore our systems. We're devastated, confused, and scared. We don't know what to do, how to fix this, or how to recover from the damage.

This nightmare scenario is all too common for online businesses today. Cyber threats are very real and can significantly disrupt operations, damage reputations, and cause major financial harm. Beyond the immediate impacts, attacks also carry regulatory, legal and human consequences.

According to an IBM report, the current average cost of a data breach is $4.24 million and it takes 287 days to identify and contain. Most are caused by malicious attacks, system glitches, and human errors. No company is immune to these risks, regardless of size or resources. Breaches and disruptions can cripple whole companies and entire supply chains if unprepared.

That's why we need to invest in comprehensive business continuity and disaster recovery (BCDR) planning. Business continuity maintains critical operations during any disruption, outage or cyberattack. Disaster recovery focuses on safely restoring systems after major incidents. Together, BCDR provides a strategic playbook to withstand and recover from technology failures, cyber breaches, natural disasters, or any crisis surpassing preventative controls. A well-prepared plan allows us to proactively address these challenges. It enables the building of resilience and response capabilities to minimize disturbances and damage when adversity inevitably arises.

By preparing for various contingencies, we can minimize potential damage caused by such events. Precautions like backup systems and processes allow us to quickly respond to and recover from cyber incidents or outages. This reduces downtime and ensures that critical operations resume as soon as possible.

In the event of a crisis, an efficient BCDR plan helps us navigate the situation while safeguarding our assets, data, reputation, and customers. This includes implementing measures to protect sensitive information and comply with regulatory requirements like data protection and privacy laws. In this way, we maintain trust with customers and stakeholders.

An efficient BCDR plan also promotes continual improvement of our preparedness against future incidents. This is achieved through regular testing, updating, and training on BCDR policies. By constantly evaluating and refining the plan, we enhance our ability to respond to and overcome potential threats.

However, creating and executing a top-notch BCDR plan is not easy. It requires carefully examining our risks, operations, dependencies, and key factors using structured frameworks. Regular risk evaluations and analysis of potential disruption scenarios are essential. This means getting leadership support and active participation from stakeholders like management, IT, security, legal, and facilities.

We must adequately fund continuity strategies, redundancies, and cyber insurance. Customized policies, procedures, and playbooks tailored to our specific risks and priorities should be developed. Regular testing of simulated cyber disasters, plan reviews, and updates are necessary to ensure BCDR effectiveness. It's also important to provide mandatory awareness, education, and skills training to employees regarding their BCDR responsibilities.

Moreover, an industry-leading BCDR plan must be constantly refined and enhanced to match the evolving risk landscape we face. As new technologies and business models emerge, new threats arise which require innovative continuity strategies. There is no one-size-fits-all, set-and-forget solution in business continuity planning. We must embed the agility to continually adapt and improve continuity measures as part of our culture.

While cybersecurity focuses on proactive defenses, unforeseen events can still disrupt critical operations. Whether cyberattacks, technology failures or natural disasters, we must prepare for outages through business continuity and disaster recovery (BCDR) planning.

BCDR provides contingencies that enable sustained functioning when facing adversities. It encompasses strategies, policies, procedures and arrangements to maintain essential processes if normal infrastructure becomes unavailable. Effective BCDR relies on understanding downtime impacts through business impact analysis (BIA), analyzing recovery priorities, and instituting resilience capabilities tailored to resource criticality.

The first step in continuity planning is systematically evaluating the potential effects of different disruption scenarios across business functions, technologies, facilities and third-party services. This "business impact analysis" (BIA) provides risk-based insights to guide strategic decisions on resource allocation and priorities for continuity investments.

BIA starts by thoroughly inventorying and documenting all business activities, the information systems and technologies supporting them, facility dependencies, supply chain relationships, and interconnections and downstream dependencies between systems. This allows us to create an impact map, visually tracing how breakdowns can propagate across upstream and downstream systems and functions due to reliance chains.

For example, a network outage could affect online sales, customer service, order processing, inventory management, shipping, and billing. A power outage could affect physical security, lighting, heating, cooling, and ventilation. A fire could affect equipment, data, documents, and personnel. A pandemic could affect staff availability, travel, and communication, while a cyberattack could affect data integrity, confidentiality, and availability.

Next, we develop threat models analyzing different disruption scenarios like network outages, ransomware attacks, fires, floods and supply chain problems. We determine cascading consequences on operations based on the mapped dependencies. We then calculate Quantified Business Impact scores for each scenario using risk analysis formulas factoring likelihood and severity. This analysis ultimately provides a prioritized roadmap guiding our continuity planning efforts.

We can use insights from the BIA to make informed decisions on creating customized continuity and redundancy strategies that match the importance of our people, processes, data and systems with the appropriate level of resilience.

For less critical systems and resources, basic data backups and eventual recovery may be sufficient, as outages will have only minor impacts. However, mission-critical revenue-generating applications require robust, redundant solutions that can maintain continuous 24/7 functionality during disruptions. Continuity strategies represent investments aligned to priorities.

There are several strategies for ensuring continuity in case of a crisis. These include having backup infrastructure and alternative sites for failover and workload mobility. We can implement emergency workarounds to sustain critical workflows during outages, and cyber insurance can provide financial protection in case of prolonged breakdowns. Multi-tiered cyber defenses, such as AI-driven threat detection and deception networks, can also be used. Regular mock disaster simulations and drills are important to validate our response plans.

To effectively implement BCDR, we need to translate analysis into actionable policies, response plans, and continuity procedures tailored to our risk profile. Leadership should establish overarching continuity policies that include governance processes, funding mechanisms, and senior oversight.

Detailed cyber incident response plans should document methods for assessing and declaring disruptions, enacting continuity contingencies, maintaining communications, and managing restoration efforts across stakeholders. Cross-functional continuity teams must be created that encompass leadership, IT, facilities, etc., and they must be equipped with response playbooks listing actions for every phase of a crisis scenario—from initial detection to long-term recovery.

We should conduct periodic plan reviews and live testing through simulated disasters across locations to validate effectiveness and uncover potential gaps. Results should feed into persistent improvement of infrastructure resilience, response processes and personnel competencies. Ongoing awareness programs should be instituted, which, much like fire drills, reinforce understanding of continuity protocols and individual responsibilities among employees.

In volatile digital environments, BCDR provides a lifeline enabling us to regain footing amidst turmoil when traditional cyber defenses falter. Just as emergency response plans engage when health deteriorates, robust continuity arrangements activate when critical infrastructure fails—automatically shifting workloads, enacting backups, and mobilizing restoration without reliance on human reaction times. This is made possible by resilient system designs that incorporate failure-handling mechanisms into digital platforms and cloudburst capabilities, providing on-demand scale during crises. BCDR offers peace of mind that critical business functions can go on despite inevitable adversity.

While continuity planning protects critical operations during outages, disaster recovery focuses on restoring platforms to normal function after major disruptions. It prioritizes the timely return of key technologies based on their business impacts, as determined through BIAs.

Disaster recovery plans detail response procedures to safely restore systems, including assessment, remediation, and testing steps. They establish RTOs and RPOs aligned to business needs that guide recovery prioritization and enhancements. Plans should cover a variety of scenarios, from isolated system failures to wide-scale events like fires, floods and malicious threats.

To minimize recovery times after isolated outages, we can use high availability (HA) configurations to remove single points of failure through internal redundancies, providing uninterrupted uptime for applications and data. We can use geographically dispersed, active-active infrastructure with automated failover to sustain operations if one location is disrupted.

Remote replication of data to distant backup facilities can be used to facilitate recovery after localized disasters. Backups should be air-gapped from primary sites to avoid a single event like malware impacting both copies. Secure cloud-based backups add resilience against disasters like fires, floods, or other destruction at physical locations.

We should conduct periodic disaster recovery testing across infrastructure tiers to validate our plans and faculties. Tabletop exercises allow us to walk through response procedures using hypothetical scenarios. Functional drills test technical failover and restoration capabilities without actual disruptions. Full-scale simulations activate plans as if an actual crisis has occurred, providing opportunities to improve strategies.

Ongoing maintenance ensures continuity arrangements stay current amid evolving business models, site footprints, technologies and threats. As priorities change, recovery strategies should be re-evaluated and realigned accordingly. Continuity considerations should be integrated into change management processes, including expansions to new locations, migrations or modernizations. Documentation should be updated whenever systems, processes or organizational structures change.

Though prevention is ideal, BCDR provides a critical safety net when disruptions inevitably strike. With strong foundations in continuity and recoverability, we can rapidly restore functionality after adverse events. We gain the agility to pivot and progress despite the uncertainties ahead. In chaotic environments, embedding resilience into our organizational DNA fosters stability for weathering ephemeral storms while advancing our long-term directions.

While technical protections are crucial for recovery, the human element is equally important. Personnel must know how to respond during contingencies and maintain operations despite infrastructure degradation. Leadership buy-in, thorough education, and clear communication readies our workforce to deliver essential services even amid strains.

We must ensure that employees thoroughly understand their continuity responsibilities and can execute disaster recovery protocols effectively through comprehensive training. Hands-on simulations offer opportunities to gain experience with response procedures in safe environments. Education should cover scenarios from minor isolated incidents to large-scale coordinated attacks.

Leadership buy-in and active promotion of BCDR establishes a culture that values and prioritizes organizational fortitude. When the C-suite demonstrates commitment by mandating planning initiatives, funding continuity teams, and participating in enterprise-wide simulations, it cascades through the entire company.

Employees become intrinsically motivated to strengthen resilience when they see executives treating BCDR as something of utmost importance rather than just a checkbox activity.

Clear crisis communication plans prepare personnel to interact appropriately within teams and with outside stakeholders during disruptions. Centralized internal channels provide timely and accurate situation updates to keep employees aligned. Meanwhile, public relations guidance shapes external messaging to maintain public confidence and mitigate reputational damage during adverse events.

Robust BCDR equips our workforce with the knowledge, experience and tools needed to maintain vital services in times of stress and uncertainty. The shared battle rhythm that comes with a subconscious understanding of response protocols enables teams to transcend hurdles. Just as military units develop cohesion and overcome adversities together through relentless drilling, organizational resilience is fostered when continuity practices permeate everyday operations and culture.

As technology becomes increasingly entwined into business functions, potential disruption vectors multiply and continuity protections must expand accordingly. Cloud adoption, remote work, mobility, and interconnected systems have dramatically broadened the risk landscapes we now face. Let's examine some key considerations for evolving our BCDR programs to match modern architectures.

For cloud-based systems, we need specialized continuity arrangements to address unique dependencies beyond our control. We should scrutinize and validate provider SLAs to guarantee adequate performance, resilience capabilities and infrastructure redundancies. We must identify alternative cloud operators for failover and factor intricacies like data residency regulations into recovery strategies. We must also institute strong cloud data governance and access controls, given the expanded exposure.

With remote employees and decentralized operations, our continuity plans should address the challenges of distributed environments. We must provide personnel with reliable mechanisms to report disruptions to centralized teams from any location. Validating remote access and cloud authentication capabilities supports extensive telework during outages, and we should consider shipping critical equipment to remote workers for local recovery when corporate facilities are impacted.

For increasingly prevalent mobile and IoT deployments, we need to account for intermittent connectivity and decentralized systems in continuity architectures. Resilient device designs, with functions like self-diagnostics and automatic failovers, can be implemented to handle intermittent network connectivity. Field personnel should be provided with hardened devices and backup communication options independent of cellular carrier networks. We should also develop procedures for coordinated physical access and repair of geo-distributed devices following disasters.

For AI/ML dependent systems, we have to consider model drift and data quality impacts during degraded operating states. We should build capabilities to detect decreasing model accuracy due to distorted data feeds after disasters, and to automatically retrain when deviations are significant.

Ultimately, business continuity considerations need to permeate our culture, policies, system designs and processes rather than being limited to siloed disciplines. When sufficiently mature, BCDR transforms from an isolated, specialized team to an organization-wide capability underpinning everything from engineering to operations. While external threats will persist, resilient entities can progress confidently, equipped with multi-layered capacities to anticipate, withstand and evolve through turbulence.

In the next chapter, we will shift from discussing contingency planning to examining effective incident response for when disasters actually strike. This entails developing orchestrated command structures, granular playbooks, crisis communication workflows, and deep forensic investigation toolsets—much like tactical response units.

Given the inevitability of some incidents surpassing our preventive barriers, we must complement proactive monitoring with extremely adept reactive capacities, directly managed by specialized teams dedicated to digital crisis response. Just as emergency first responders have sophisticated protocols, cyber resilience requires dedicated defenders continually honing response tactics and tools. We will explore what it takes to establish a world-class security operations center that can proactively defend critical infrastructure.

Chapter 9

INCIDENT RESPONSE VS. INCIDENT COMMAND

Business continuity and disaster recovery (BCDR) plans are essential components of a comprehensive cybersecurity strategy, serving as reliable shields that protect against expected hazards. However, simply having these plans in place is not enough to ensure full protection. In the rapidly evolving digital landscape, cyberattacks can easily bypass traditional defenses, requiring a different kind of resilience that is built through swift and effective incident response.

In this chapter, we will delve into the heart of this critical response mechanism, exploring the roles of incident response and incident command in defending against the digital storms that threaten organizations today. By understanding and implementing robust response capabilities, companies can enhance their cybersecurity posture and minimize the impact of cyber incidents on their operations and reputation.

Using my experience as a certified incident commander at the Department of Homeland Security and a seasoned fighter against ransomware attacks, I will reveal the inner workings of incident response. We will explore the technical expertise, strategic coordination, and collaborative efforts that turn chaotic moments into opportunities for resilience.

Additionally, my long-standing partnership with the Federal Bureau of Investigation, developed through the FBI CISO Academy, will provide unique insights into how private and public expertise unite to combat cyber threats.

Imagine having a meticulously crafted BCDR plan in place, with strong firewalls and secure backups. However, a clever cyberattack manages to penetrate the defenses, throwing your organization into chaos. Without a solid incident response framework, this situation can quickly lead to operational paralysis and reputational harm. This is where incident response and incident command come into play, working together seamlessly to contain the damage, eliminate the threat and facilitate a swift recovery.

My personal experience, particularly in the face of ransomware attacks, has reinforced my belief in the effectiveness of a well-coordinated response. Witnessing dedicated teams smoothly transitioning from pre-planned BCDR protocols to real-time incident response has emphasized the importance of adaptability, agility and technical expertise in the ever-growing landscape of threats.

An organization's cybersecurity strategy relies heavily on incident response, which serves as the final barrier when all other security measures have failed. A well-structured incident response framework allows organizations to reduce the consequences of cyberattacks, minimize operational disruptions and safeguard their reputations.

The effectiveness of incident response lies not in rigid checklists, but rather in utilizing a combination of skills. It's similar to being the first one to act and solve the problem, or the sharp analyst who can interpret digital clues. It's like being on a team that swiftly transitions from initial detection of the issue to containing and eradicating it.

The first line of defense is the swift detection of anomalies. This requires advanced monitoring systems, skilled analysts who can interpret subtle changes in network behavior, and a culture of vigilance within the organization. These watchful eyes are invaluable in spotting the initial signs of a cyber wildfire.

However, detection is just the first step. To successfully respond to a cyber incident, a comprehensive investigation is necessary to uncover the attack vectors, identify compromised systems, and understand the adversary's motives. This critical knowledge enables the incident response team to prioritize containment efforts and develop precise mitigation strategies, much like a skilled surgeon locates the source of a patient's pain before performing a procedure.

By thoroughly investigating the incident and gathering essential information, the team can make informed decisions and take targeted actions to contain the threat and prevent further damage to the organization's systems and data. This proactive and meticulous approach to incident response is crucial for minimizing the impact of cyber incidents and ensuring a swift, effective recovery.

Clear, concise communication is also necessary to ensure swift and coordinated action across all levels of the organization. Keeping leadership, technical teams, and affected departments informed at every step builds trust, minimizes disruption and facilitates a unified operation.

This ascertains that everyone knows their role in extinguishing the digital flames.

The ultimate goals are to contain damage and eliminate threats. This may involve isolating infected systems, shutting down vulnerable services and deploying specialized tools to neutralize malware. It's a delicate dance of precision and speed, similar to a firefighter isolating a burning building to prevent the fire from spreading.

However, the job doesn't stop when the immediate crisis is over. Evidence must be gathered and data analyzed in order to prevent future attacks. Forensic analysis involves meticulously collecting and analyzing log files, network traffic data and system artifacts. This process provides valuable insights into the techniques and motives of the adversaries. Armed with this information, organizations can strengthen their defenses and actively search for future threats, using it as their shield and sword.

In addition to technical measures, it is essential to consider the broader organizational impact of cyber incidents when developing a thorough defense strategy. Effective incident response requires more than just technical expertise—it demands collaboration with stakeholders across the organization, including legal, public relations, and other departments. A holistic approach that incorporates legal, reputational, and other considerations is necessary for organizations to not only survive cyber incidents, but also emerge from them stronger and more resilient.

This line of planning acknowledges that the impact of a cyberattack goes beyond the technical realm, and that a coordinated, cross-functional reaction is crucial for mitigating the full range of potential consequences. Through working together and considering all relevant factors, organizations can enhance their ability to respond to and recover from cyber incidents, ultimately building stronger defenses and a more robust cybersecurity posture.

In today's fast-paced digital age, technology continues to evolve at an unprecedented rate. This rapid advancement has brought many benefits, but it has also enabled cyber adversaries to develop increasingly diverse and sophisticated methods. As a result, the traditional style of incident response, which relied solely on IT departments to identify and address security breaches, is no longer sufficient. The constantly changing landscape of cyber threats requires broader, more adaptive planning that accounts for the complex and progressive nature of technology.

By recognizing the pace of technological change and the challenges it poses, organizations can develop incident response strategies that are better equipped to handle today's dynamic, unpredictable cyber threats. Instead of simply reacting to incidents as they happen, they can anticipate and circumvent potential dangers before they even materialize. This comprehensive approach involves continuously monitoring and analyzing the organization's systems and networks, identifying vulnerabilities, and implementing appropriate security measures. Moreover, by combining these functions, organizations can gain a better understanding of their overall risk profile and the potential impact of cyberattacks on their operations.

Efficient incident response mechanisms, along with specialized teams and established frameworks, are crucial for maintaining cyber resilience. The Incident Command System (ICS) provides a strong, structured defense foundation—a standardized roadmap to navigate through an attack. By following organized protocols for assessment, decision-making and communication, the incident response becomes a highly coordinated effort led by experts, transforming it from chaos into coherence.

Playbooks, which record collective wisdom and best practices, are vital in promoting systematic crisis management and setting up productive countermeasures. Like well-rehearsed fire drills, they minimize confusion and maximize the overall effectiveness of the operation.

A clearly defined incident response framework serves as the backbone of organizational resilience against cyber threats. It acts as a strategic guide, outlining the systematic steps to be taken during a security incident. This framework not only streamlines the response process but also mitigates potential damage by facilitating swift and coordinated actions.

Structured protocols within the framework guarantee a methodical and expedient reaction. Specialized incident response teams, equipped with the knowledge and expertise to implement these protocols, become the frontline defenders. Their harmonized efforts enhance the organization's ability to detect, contain, eradicate and recover from security incidents.

But this approach offers more than immediate incident handling. It enables organizations to learn from each incident, continuously improving and adapting their strategies. This iterative process reinforces their overall security posture, making them more resistant to future cyber threats.

In essence, a well-defined incident response framework is not just a set of guidelines; it's a strategic asset that empowers organizations to face and mitigate the ever-evolving landscape of cybersecurity challenges. It's also a critical component in building cyber resilience—a blueprint by which organizations can effectively counter security incidents, ensuring that the right actions are taken at the right time.

The future of incident response lies in more than just reacting to breaches. It's about taking a proactive approach, like a skilled hunter tracking their prey. Embracing technical expertise and structured workflows allows organizations to tackle crises head-on. This means having intensive skills in threat intelligence, vulnerability scanning and malware analysis to detect anomalies before they become major incidents. Continuous monitoring and threat hunting become the preemptive strike, stopping digital wildfires before they even start.

Through my experience in managing recovery efforts for ransomware attacks, I've seen the importance of this proactive shift. By anticipating threats, implementing robust monitoring systems and constantly refining defensive strategies, organizations can greatly reduce their vulnerability to cyberattacks. It's the same as taking care of your health by identifying predispositions before they turn into full-blown ailments.

When crises do eventually arise, effective incident response teams (IRTs) are the backbone of the solution. Made up of security professionals with diverse skill sets, IRTs are the ones on the ground, turning strategic plans into actionable steps. They follow incident response frameworks, collaborate closely with the incident command center, and handle the technical complexities during active incidents. It's a combination of proficiencies, with analysts deciphering digital clues, engineers utilizing technical tools, and communicators keeping everyone informed. All of this is orchestrated by the IRT to efficiently resolve the situation.

Serving in the field, I have several times seen IRTs' ability to smoothly transition from pre-planned protocols to real-time incident response, as well as their agility in adapting to the ever-changing nature of the cyber threats at hand. These skills are crucial for the swift and effective resolution of crises. Witnessing firsthand the dedication and skill of IRT members in tackling cyberattacks has cemented my belief in their importance as the frontline of organizational defense.

While incident response, incident command and IRTs each play a vital role, it is their excellent collaboration that unlocks the full potential of cybersecurity. These three elements are like a well-rehearsed orchestra, with each instrument contributing its own unique melody to create a harmonious outcome. Incident response provides the strategic vision; incident command orchestrates the overall operation; and IRTs execute the technical maneuvers on the ground. This synergy, this unified chorus of expertise, is what transforms potential disasters into opportunities for growth and resilience.

Some time ago, within the FBI CISO Academy, a group of professionals from both the private and public sectors came together for an intense incident response team (IRT) simulation. Their mission was to combat a sophisticated ransomware attack that had targeted a critical infrastructure network, demanding immediate action.

In this high-pressure situation, the private sector experts, who were well-versed in threat intelligence and malware analysis, seamlessly joined forces with their public sector counterparts. The representatives of the public sector brought their regulatory insights and legal perspectives to the table, creating a powerful partnership.

The joint IRT faced the challenge of quickly containing the ransomware, assessing the impact on vital systems, and formulating a comprehensive response strategy. The private sector cybersecurity analysts deployed advanced tools to identify the malware's origins and propagation methods, while those from the public sector navigated regulatory frameworks and legal considerations for a lawful and coordinated campaign.

This endeavor not only showcased the diverse skill sets within the IRT, but also underscored the importance of harmonizing private and public sector expertise. It highlighted the blend of technical acumen, legal compliance, and regulatory awareness required to effectively manage the evolving landscape of cyber threats. In the digital battlefield, just as in the real world, unity and collaboration are the ultimate weapons against adversity.

With the key players of incident response—frameworks, teams, and proactive strategies—firmly established, let's delve into a specific framework that is causing a stir in the field of cybersecurity: the Incident Command System (ICS). Traditionally, ICS has been a stalwart ally in the physical world, aiding first responders in battling wildfires or coordinating relief efforts during earthquakes. Now, it has expanded its support into cybersecurity.

Imagine the DHS's First Responder ICS as a seasoned warrior, battle-tested and adaptable, now lending its prowess to the cyber world. Adapting ICS for cybersecurity goes beyond mere terminology importation; it involves harnessing its core principles to navigate the unique landscape of digital crises. It's like translating a map designed for mountains to the slopes and hidden ravines of a different terrain.

This translation means embracing ICS as a standardized, all-hazard incident management concept that is equally as applicable online as it is to physical emergencies. ICS provides a common language, facilitating seamless collaboration between diverse teams. Envision security analysts, IT specialists and legal professionals, all working under the unified banner of incident command.

ICS's focus on minimizing damage, reducing recovery time, and optimizing resource allocation becomes even more crucial in the fast-paced field of cyberattacks. Every minute counts and every decision can tip the scales. ICS offers a structured outline for swift assessment, clear decision-making, and efficient resource deployment, ensuring the team operates with calculated precision rather than scrambling in the dark.

But ICS isn't just about rigid protocols; it recognizes the vital role of technology, processes, and people in crafting a thorough incident response. It allows organizations to make the most of their existing tools and expertise while providing a framework for smoothly combining them. With ICS, teams can move past reactive firefighting and focus on strategic countermeasures, staying one step ahead of adversaries.

This brings us back to the heart of incident response: the swift detection and containment of breaches. ICS, when adapted for cybersecurity, becomes a force multiplier, amplifying the effectiveness of incident response teams. It ensures every technical resource is utilized optimally, every decision made minimizes potential damage, and no minute spent investigating is wasted.

In essence, ICS bridges the gap between the physical and digital realms, offering a battle-tested framework for navigating the complexities of cyber incidents. By translating its core principles and adapting them to this new field, organizations can harness the power of ICS to build a stronger, more proactive, and ultimately victorious response to the changeable threats of the cyber age.

As we bring this exploration of incident response and incident command to a close, recall the captivating image of an orchestra, with each instrument playing its part in a harmonious reply to the digital storms of our time. Incident response, led by skilled analysts and proactive hunters, takes on the melody, identifying threats before they can cause chaos. Incident command, like a conductor, coordinates the overall program, ensuring that every note is perfectly executed. The dedicated members of incident response teams, known as instrumentalists, execute the technical maneuvers with precision and expertise.

The shift towards more orchestrated cyber defense represents a significant advancement in the fight against cyber threats. While we are making progress in moving away from the reactive firefighting of the past, it's important to recognize that the cybersecurity landscape remains dynamic and complex. Although our focus has shifted towards proactive measures, anticipating dangers, and building resilience, we must remain vigilant and adaptable as new challenges emerge. By acknowledging both the progress made and the work that lies ahead, we can continue to move towards a safer and more secure digital world.

While in my current role as Field CISO, I have personally witnessed this shift through countless battles against ransomware and the subsequent recovery efforts. It is fueled by a powerful combination of technical ability, structured frameworks like ICS, and unwavering collaboration. By leveraging the knowledge and resources of organizations like the FBI, through initiatives such as the CISO Academy, we strengthen our collective defenses and elevate our individual capabilities. Together, we form a united front, working in harmony to anticipate, detect, and respond to constantly evolving cyber threats.

Our journey has taken us from the heart of incident response, where technical expertise confronts digital threats, to the broader stage of incident command, where strategic decisions guide the course of action. We have explored the crucial role of frameworks like ICS in providing structure and coherence, and we have celebrated the power of proactive threat hunting and data analysis.

Ultimately, it's the human element that truly ignites the engine of cyber resilience. The dedication of incident response teams, the leadership of incident commanders, and the collaborative spirit of the broader cybersecurity community are the forces that will propel us forward. As we continue to hone our skills, refine our frameworks, and deepen our shared bonds, we can confidently face the cyber storms that lie ahead, emerging stronger, more resilient, and ever more determined to safeguard our digital frontiers.

Chapter 10

BREACH COACH

The memories of my first major breach response call still haunt me to this day. As a new coach at the time, I was thrown into the tumult of deciphering complex log files and frantically attempting to locate the malware that had infected a healthcare company's network. For several days, I was always on edge due to the adrenaline rush that came with creating containment tactics. When I finally saw operations returning to normal, a sense of relief came over me. What had begun as just another job quickly evolved into a true calling: guiding organizations through the chaos of a cyber breach and back to stability.

Imagine the aftermath of a cyberattack: data vanishing into the unknown, business operations screeching to a halt, and anxiety running rampant within the organization. Navigating these cyber storms requires more than just a theoretical understanding of incident response; it necessitates developing the skill to excel under immense pressure.

For those involved in containment, outlining strategies must become second nature, much like a seasoned sailor instinctively adjusting the sails to counter shifting winds. Each incident response is like a treacherous sea voyage, demanding both technical expertise and the ability to stay calm amid the storm. While there may be no foolproof map to guide the journey, working with an experienced navigator who can steer through even the most challenging situations becomes a priceless asset in containing a breach and restoring stability to the organization.

This is precisely why breach coaches have become invaluable across industries. According to a recent Forrester survey, 55% of firms are increasing their budgets for third-party incident response support. This highlights the growing recognition of the benefits of an impartial coach with insider knowledge. Just as a skilled surgeon can swiftly address a severe injury, an experienced breach coach can significantly reduce crisis response time, potentially shaving off precious hours. While there is no set timeline for a successful breach response, I have led teams that successfully mitigated advanced ransomware attacks in under eight hours—nearly three times faster than the average time without specialized coordination.

The duration of a response depends on various factors, such as the organization's cyber posture before the attack, the experience and expertise of the response team, and the nature of the breach itself. However, in the high-stakes world of cyber breaches, one thing remains clear: when data is leaking rapidly, every second counts, and a rapid counteraction is as critical as the swift action of a skilled surgeon in the operating room.

Gone are the days when breach coaches were a luxury reserved for only the largest corporations. In today's complex cyber threat landscape, where even small businesses hold valuable data, breach coaches have become an essential component of robust security. They are not a mere checkbox to be ticked off, but rather a proactive investment in resilience, minimizing the damage caused by cyberattacks.

In this chapter, we delve deep into the vital role of breach coaches, elucidating their responsibilities and underscoring the concrete benefits they offer. Consider them the cybersecurity equivalent of an elite SWAT team, adeptly coordinating response efforts, mitigating risks, and leading organizations back to stability. As you read on, picture that calm presence in the midst of chaos, and discover how breach coaches can become the beacon of hope for your organization during the most challenging moments of an attack.

When your defenses are breached by a cyberattack, the path ahead may appear uncertain and unclear. This is where the breach coach steps in as your reliable compass, a seasoned traveler through the treacherous landscape of incident response. Their responsibilities are vast, requiring not only technical expertise but also leadership, effective communication and empathy.

At the core of their role lies guiding your organization through the immediate aftermath of an attack. They become your crisis managers, assisting you in navigating the initial shock, containing the breach, and assessing the extent of the damage. But they also comprehend the human toll of such attacks, guiding your leadership through challenging decisions and maintaining transparent, consistent communication with stakeholders, customers and the public.

Rather than mere firefighters extinguishing the flames, think of breach coaches as architects rebuilding your security infrastructure. They work hand in hand with your incident response team, offering expert guidance on containment, eradication, and remediation strategies. However, their value extends beyond the technical realm. They act as liaisons between your organization, law enforcement agencies and other external parties, ensuring seamless collaboration and adherence to legal requirements.

Remember, a breach coach is not a magical solution, but a force multiplier. They enhance your existing defenses, augment your team's skills, and guarantee a swift, efficient, and ultimately successful response to a cyberattack.

Hiring a breach coach isn't simply adding another piece to your security puzzle; it's like acquiring a Swiss Army knife, packed with tools that unlock an array of benefits. Imagine the panic and confusion that follow a cyberattack. Time becomes your enemy, and every decision becomes critical. A breach coach becomes your timekeeper, streamlining response efforts and allowing for swift, decisive action. Their expertise translates into measurable improvements in efficiency, reducing downtime and minimizing the financial impact of the attack.

But their advantages go beyond these. Consider the immense responsibility weighing on leadership. Breach coaches become their trusted advisors, offering wise counsel and relieving the burden of decision-making. They navigate the emotional storm alongside your team, minimizing stress and anxiety and ensuring everyone remains focused on the recovery task at hand.

Breach coaches also offer an objective perspective, challenging assumptions and encouraging innovative thinking. They create a safe space for the team to express their emotions and concerns, enabling them to work through issues and find solutions together.

Think of them as translators as well, bridging the gap between technical jargon and plain English. They facilitate clear, concise communication throughout the organization, keeping everyone informed and aligned. This transparency fosters trust and teamwork, which are crucial elements in successfully weathering the cyber storm.

Do you remember the shattered remnants of an organization's reputation after an attack? Well, breach coaches, with their skills in crisis communication, become your architects of resilience. They help you craft a clear, honest narrative, minimizing reputational damage and rebuilding public trust. With their guidance, you will emerge from the crisis stronger, wiser, and better prepared to face future threats.

A breach coach is a steadfast partner in restoring business health, public trust, and, most importantly, human dignity. It is crucial for every breach response plan to assign a coach who can provide emotional support and stability. Studies show that organizations with dedicated coaches report 25% less leadership burnout post-incident.

With so many options out there, finding the perfect breach coach can feel like walking through a minefield. But fear not, because we've got you covered. Let's dive into the key factors you should consider when choosing the ideal coach for your organization.

First and foremost, experience is crucial. Look for someone who has a proven track record of successfully handling incidents across various industries and threat scenarios. Their battle scars are a testament to their ability to navigate even the most treacherous terrain.

However, experience alone is not sufficient. It's important to find a coach who not only speaks your language technically but also aligns with your organization's values. Can they integrate cleanly into your team dynamics and build trust with your leadership? Finding a cultural fit is like discovering the perfect key that unlocks strong collaboration and mutual respect.

When you embark on your search for a breach coach, it's important to keep these considerations in mind. They will help you find the perfect guide to navigate the darkest nights of a cyberattack. The term "breach coach" may evoke a certain image, but witnessing their expertise in action takes your understanding to a whole new level. That's where case studies come into play—they provide windows into past battles, showcasing how breach coaches have successfully steered organizations through the chaos of cyberattacks.

Let's imagine a scenario where a small healthcare provider is hit by a ransomware attack, with patient data hanging in the balance. The breach coach steps in, exuding a calm demeanor that soothes the panic. They operate with the precision of a master chef, skillfully coordinating the various components of the response—containment, negotiation, and recovery—into a cohesive and effective resolution. In the aftermath, data is restored, operations resume, and the public commends the organization's resilience. This is the tangible impact a coach makes in leading organizations through the complexities of a breach.

But the stories of breach coaches go beyond successful resolutions. Imagine a multinational corporation grappling with a data breach while international media circles like a vulture. A skilled breach coach steps in and guides their communication strategy, crafting a transparent, honest story that rebuilds trust and minimizes reputational damage. This too is a victory, showcasing the power of a breach coach in navigating the choppy waters of crisis communication.

These examples are just glimpses into the many breach coach success stories. As you delve deeper into case studies, you'll encounter tales of diverse threats, varied industries, and unique challenges overcome. These stories are not just data points; they are compelling narratives that highlight the power of having a breach coach by your side.

The cyber battlefield is constantly shifting, and many dangers lurk in the digital shadows. However, breach coaching is evolving at an even faster pace, always ready to face these emerging challenges head-on.

Think of it as an arms race, where cybercriminals develop advanced weapons and breach coaches refine their defenses. We are witnessing the rise of specialized experts in niche areas like cloud security and AI-powered attacks. They are the digital knights, prepared to take on any dragon that the threat landscape throws their way.

Technology serves as a powerful ally in this ongoing battle. Picture breach coaches armed with advanced tools that enable them to automate forensic analysis, facilitate seamless communication, and predict adversary behavior. These technological innovations act as their armor and arsenal, providing the resources they need to swiftly and accurately respond to ever-evolving threats. As cyberattacks grow more complex, these cutting-edge tools become indispensable in the hands of skilled breach coaches, enhancing their capabilities and fortifying their defenses against the relentless onslaught of cyber adversaries.

However, amidst all the technological progress, one thing remains constant: the human element. The future of breach coaching lies in fostering empathy, building trust, and inspiring resilience within organizations. These coaches are more than just technical geniuses; they're crisis counselors, strategists, and allies in the fight against cybercrime.

So, as we look ahead to the future of cybersecurity, let us remember this: the role of the breach coach will only grow more vital. They will be the silent guardians, the unwavering anchors, and the guiding lights that navigate us through the ever-changing tides of the digital world.

When it comes to breach coaching, the famous quote "With great power comes great responsibility" couldn't be more accurate. Dealing with the aftermath of a cyberattack often presents ethical dilemmas that demand careful consideration. Imagine yourself at a crossroads—one path tempting you with quick solutions while the other is lined with ethical principles. A good breach coach must possess a strong moral compass to direct them through this complex maze.

One of the biggest challenges lies in potential conflicts of interest. What if a law enforcement agency investigating the attack requires access to confidential information held by the coach? Striking a balance between upholding legal obligations and maintaining client confidentiality is a delicate issue, requiring clear communication and transparent protocols.

Another ethical minefield lies in the area of disclosure. How much information should be shared with stakeholders, the public, and even within the organization itself? It's crucial to achieve the right balance between transparency and protecting sensitive data, and a breach coach must be a master of conveying clear, truthful and compelling narratives without jeopardizing individual or organizational security.

However, ethical dilemmas are not just about avoiding pitfalls; they also present opportunities for positive influence. A breach coach can also use their expertise to advocate for stronger data privacy regulations, or push for increased cybersecurity awareness within local communities. Their voice can become a powerful force for good, shaping a more secure digital landscape for everyone.

Remember, ethical considerations are not limitations that hinder a breach coach's effectiveness. They serve as guiding lights, ensuring the coach's actions are just and responsible as well as successful.

Engaging the services of a breach coach may appear to be a reactive measure—a response to a perceived vulnerability. However, it encompasses much more than that. It is a proactive investment in resilience, showing your dedication to safeguarding your most valuable assets in an increasingly digital world.

Again, think of breach coaches as skilled architects constructing robust firewalls and sprinkler systems, instead of mere firefighters who arrive after the flames have caused havoc. They empower your organization with the knowledge and composure to withstand the scorching heat of a cyberattack and emerge even stronger on the other side.

Their value extends beyond immediate crisis response. They become allies in your ongoing cybersecurity journey, mentors who train your team, partners who refine your protocols, and advocates who champion a culture of security throughout your organization.

In an era where a single click can trigger a digital storm, having a skilled, ethical breach coach by your side is no longer a luxury; it's a necessity. They will be your cyber sentinels, standing guard against unseen threats, ready to guide you through the dark night of a cyberattack and lead you toward a dawn of resilience and recovery.

Remember, the choice to invest in a breach coach is not simply about protecting your digital assets; it's about protecting the trust of your customers, the well-being of your employees, and the future of your organization.

Chapter 11

CYBER COUNSEL

When a cyberattack strikes, the battleground shifts from the digital realm to the legal arena. A breach leaves behind a chaotic landscape of shattered data, giving rise to a multitude of legal challenges that pose a threat to the organization. Successfully navigating this treacherous terrain requires more than just solid security measures; it demands the expertise of a seasoned guide and a cyber legal expert to lead the way through the storm.

In our previous conversations, we have highlighted the significance of breach coaches in navigating crises. Yet, we must acknowledge that they constitute just one component of a comprehensive cybersecurity strategy and effective crisis management. Cyber legal experts form another integral part of this equation, working alongside breach coaches and technical teams to provide strategic guidance. While breach coaches focus on incident response, legal experts collaborate across departments during both calm and tumultuous times, helping leadership strike a delicate balance between security imperatives and business goals.

Consider the following scenario: You awaken to the shocking revelation that your organization, previously believed to be a fortress of digital security, has been devastated by a cyberattack, leaving behind a charred landscape of exposed data and compromised confidential information. Headlines blare news of your company's downfall, while a storm of lawsuits loom on the horizon, threatening to hasten your demise. This bleak picture illustrates the consequences of an unprepared cyberattack—a legal quagmire where even a single misstep can plunge an organization into irreversible ruin.

Even if your organization is equipped with state-of-the-art security software and robust safeguards, it is imperative not to underestimate the legal ramifications of a breach. Neglecting this aspect can plunge your leadership into an even more chilling abyss than the initial cyberattack. Although the technical damage can be mitigated and repaired, the complicated legal challenges that follow it possess the power to bring an entire company to its knees.

As we saw in earlier chapters, the 2017 Equifax breach serves as a grim reminder of this reality. It was like an earthquake that shook the corporate world to its core. The foundation beneath Equifax, a trusted credit reporting agency in America, crumbled, exposing the personal data of a staggering 147 million individuals. Names, Social Security numbers, birth dates, and even driver's licenses were laid bare for malicious actors to exploit. This was not just a data breach; it was a brutal revelation of the gap between digital security and legal preparedness.

The consequences were swift and merciless. Regulatory bodies like the Federal Trade Commission (FTC) swooped in, questioning Equifax's inadequate security measures and delayed disclosure. A flood of class-action lawsuits followed, accusing the company of negligence and seeking compensation for the exposed data. Ultimately, Equifax paid a record-breaking $700 million settlement as the price for their failure to take responsibility.

The financial burden was just the beginning, as Equifax's once-trusted brand became synonymous with vulnerability. Overnight, consumer trust that had been built over decades disappeared. This event serves as a stark reminder of the high cost of neglecting the legal aspects of cybersecurity. The Equifax breach was a catalyst for change, ushering in a new era where legal expertise actively confronts cyber threats.

In today's ever-changing world, legal knowledge has become a crucial weapon in the fight against cybercrime. Firewalls and other technical measures alone are not enough to protect against cyber threats and the consequences of a breach. This has given rise to a new breed of professionals known as cyber counsel, who bridge the gap between law and technology within organizations.

Cyber counsels are not your typical courtroom attorneys; rather, they serve as digital diplomats, adeptly navigating the intricate web of data breach notification laws, international regulations, and potential litigation risks. Their crisis management skills and strategic communication plans complement the work of breach coaches, who manage media storms and address public concerns.

To foster digital resilience, cyber counsel identifies vulnerabilities and guides leadership in striking a balance between robust security measures and business objectives. It is essential to involve legal teams early in the planning process, as they work closely with breach coaches to ensure a cohesive and comprehensive response to cyber incidents.

This collaboration helps maintain a consistent approach and reinforces an organization's ability to successfully weather the storm of a cyberattack.

Every organization has its own unique environment when it comes to cybersecurity. With different data, operations, and risk profiles, it's crucial to understand the diverse roles and responsibilities of cyber counsel. No longer does legal support only come into play after a cyberattack. Today, cyber counsel operates on a whole new level, equipped with a versatile skill set to tackle specific threats and challenges in the digital landscape.

One of the crucial tools in their arsenal is deciphering the complex web of data breach notification laws. Each region and nation has its own set of regulations, making it a daunting maze to navigate. Cyber legal experts act as translators, ensuring that your organization complies with all relevant obligations, from prompt notification procedures to mitigating potential fines and lawsuits.

Take the example of Marriott, a hotel giant that suffered a massive breach impacting millions of guests worldwide. Their cyber counsel had to quickly identify and comply with over 60 different breach notice laws across various states, nations, and international jurisdictions. From determining who to notify based on local requirements to meeting tight reporting deadlines, their legal team played a crucial role in preventing regulatory non-compliance consequences.

But while compliance is a critical foundation, it is only the beginning. In the wake of a successful cyberattack, organizations often face a deluge of media inquiries and public scrutiny. This is where cyber counsel, serving as strategic communication experts, play an essential role in partnership with breach coaches. By working together, they craft coherent and transparent messages that soften reputational damage and prevent misinformation from spreading. As the storm of public attention intensifies, this united front of legal and coaching expertise forms a bulwark against panic, keeping your communications consistent, accurate, and reassuring throughout the crisis.

Yet, their expertise goes beyond just handling immediate crises. Cyber counsel are proactive experts in digital resilience. They conduct legal assessments of your security policies, identifying vulnerabilities before malicious actors can exploit them. They provide guidance on incident response procedures, ensuring that evidence is preserved and legal risks are minimized. In today's ever-evolving threat landscape, they stay one step ahead, constantly learning about emerging legal challenges such as AI-powered attacks and quantum computing. This future-proofs your defenses, making you ready to face any new threats.

What sets cyber legal experts apart is their unique blend of skills. They are more than just lawyers or crisis heroes; they are legal minds who also speak the language of technology. They understand the complexities of firewalls and encryption protocols. This allows them to bridge the gap between IT and legal teams, fostering seamless collaboration and ensuring that your cybersecurity posture is truly holistic.

In short, the strength of cyber counsel lies in their ability to combine legal expertise with technical understanding. With their support, your organization not only survives cyberattacks but emerges stronger, more resilient, and legally well-protected.

With such formidable allies available for cyber defense, how can you find the perfect one for your organization? Having the right cyber counsel is invaluable when it comes to navigating the legal aspects of cybersecurity. It's not about selecting just anyone to play the role—understanding your options and finding the ideal fit is essential in building a strong digital defense.

For organizations with abundant resources, an in-house cyber counsel specialist can offer unparalleled familiarity with your company's own culture and needs. They are readily available, integrated into your internal processes, and possess a deep understanding of your risk landscape. However, it's important to consider the limitations, such as workload constraints, a potentially narrower field of knowledge, and the risk of becoming too accustomed to internal biases.

Partnering with an external law firm opens doors to a wider pool of specialized talent. You gain access to diverse legal expertise across data privacy, cybercrime, and specific regulations, ensuring comprehensive coverage from all angles. On the other hand, external lawyers may lack in-depth knowledge of your personal business intricacies, and communication channels might be less streamlined, potentially affecting response times.

Strategic partnerships are another good option. By building relationships with legal partners in specialized niches like data privacy or international regulations, you can tap into targeted expertise when needed. This approach allows you to access a broader network of knowledge while maintaining cost-effectiveness.

Additionally, cybersecurity insurance with bundled legal assistance can provide a surprisingly flexible solution. Many insurance providers now offer legal support services as part of their packages, granting immediate access to qualified cyber counselors during an attack. Although this option may have limitations on the scope and depth of legal services, it presents a cost-effective way to ensure protection.

The key to building the perfect cyber counsel team lies in tailoring it to your organization's specific needs. Take into account factors such as budget, industry regulations, and the complexity of your IT infrastructure when making your decision. Don't be afraid to mix and match different options to create a hybrid team that combines in-house expertise with external specialists for maximum impact.

When partnering with cyber counsel, whether they are in-house or external, best practices must be followed to ensure effective collaboration during a cyberattack. Firstly, you must clearly define the roles and responsibilities of the cyber counsel within your incident response plan. Make sure everyone understands their specific duties in terms of communication, legal risk assessment, strategic advisory, and integration with technical teams.

Secondly, establish organized intake processes and channels that are optimized for legal teams to receive timely and accurate updates during crises. Speed and transparency of information flow to counsel are of utmost importance.

Thirdly, maintain open pathways for counsel to continuously brief leadership and technical staff with informed legal assessments, projections, and advice throughout incidents. This will enable optimal decision-making.

In addition, create quick reference materials that distill counsel guidance into simplified action steps that non-legal staff can easily implement during a response. This will ensure the smooth execution of best practices.

Lastly, involve counsel proactively in crisis simulations and preparedness exercises to build institutional readiness and muscle memory well before real incidents occur. This will embed a sense of responsiveness within your organization.

Investing in a strong cyber counsel team is like fortifying your organization's defenses against legal adversaries in the digital realm. These experts provide both legal assistance during a crisis and strategic guidance throughout the entire cybersecurity lifecycle. By having this capability in place, you are preparing yourself for any adversity that may come your way.

Take the example of the US retailer Target, which fell victim to a massive data breach that exposed the payment card details of millions of customers. This cyberattack was one of the largest ever experienced by a private company.

When the breach was discovered, Target's cyber counsel sprang into action. They collaborated closely with the technical teams to identify how the attack occurred, determine the affected systems, and assess the extent of the damage. At the same time, they carefully analyzed the legal obligations surrounding breach notification and reporting, as dictated by state and federal laws.

The counsel provided valuable advice to the organization's senior leadership, informing them about potential legal liabilities, penalties, and the looming threat of lawsuits. They also played a crucial role in devising a clear communication plan that would transparently inform customers and stakeholders without exacerbating the legal risks. When regulators and law enforcement agencies arrived to investigate, the counsel skillfully managed the exchange of highly sensitive information while ensuring full cooperation.

In the aftermath, Target faced over 90 lawsuits, including a massive class action suit. However, thanks to the expertise of their cyber counsel, the majority of the consumer class action claims were dismissed using arguments rooted in data security law. Furthermore, during the negotiation of the final settlement, the counsel was able to significantly reduce the potential damages.

Despite the significant financial losses incurred by the breach, Target's swift response and engagement of cyber counsel were instrumental in navigating the complex legal landscape and minimizing long-term reputational and regulatory consequences. This case study underscores the essential role of cyber counsel in steering organizations through the tumultuous aftermath of a cyberattack, as well as their value in mitigating any legal fallout and fostering recovery.

By allocating resources to build a strong cyber counsel team, you are not only bolstering your organization's long-term resilience but also securing a sense of reassurance. These skilled professionals serve as your vanguard in confronting the inevitable legal complexities arising from the dynamic and rapidly changing world of cybersecurity.

In the face of a cyberattack, your organization may be confronted with a multitude of different legal challenges. Successfully navigating this storm and effectively addressing subsequent obstacles necessitates a comprehensive understanding of various critical domains.

One crucial area is data privacy and protection laws. It is imperative to comprehend the legal frameworks and regulations governing the safeguarding of sensitive data, such as personally identifiable information (PII) and trade secrets. Compliance with regulations like the General Data Protection Regulation (GDPR) in the European Union or the California Consumer Privacy Act (CCPA) in the United States is indispensable to avoid penalties and maintain the trust of customers and stakeholders in the aftermath of a breach.

Additionally, your organization must be acquainted with cybersecurity laws and regulations specific to its jurisdiction. These laws not only outline the necessary security measures but also mandate reporting cyber incidents to authorities and conducting routine security audits. By staying abreast of legal requirements, you can proactively implement robust security practices and lessen the risk of attacks.

When a cyberattack occurs, organizations bear the responsibility of notifying affected individuals, regulators, and other relevant parties. Understanding the breach notification obligations within the given jurisdiction is necessary to prevent penalties and safeguard reputation.

Organizations must adhere to specified notification timelines and follow proper procedures to ensure transparency and effective communication during the aftermath of a cyber incident.

Contracts with partners, clients or vendors also play a significant role in managing the legal aftermath of a cyberattack. It is crucial that your organization carefully assesses its contractual obligations and liabilities, particularly those related to cybersecurity and data breaches. In such instances, adherence to indemnification clauses, limitations of liability, and other relevant aspects are vital to mitigate legal ramifications and potential disputes.

Furthermore, cyberattacks often result in the theft or compromise of intellectual property, making it paramount for organizations to understand the legal landscape surrounding intellectual property rights. To protect their assets, companies should consider implementing copyright protection measures, securing patents, or even taking legal action against the perpetrators. By actively safeguarding their intellectual property, organizations can minimize the negative impact of cyber theft.

Another important aspect to consider is insurance coverage for cybersecurity incidents. Organizations need to fully understand the terms, conditions, and coverage of their insurance policies in order to effectively seek compensation for damages caused by cyberattacks. Comprehensive insurance coverage can help alleviate financial losses and expedite the recovery process.

In addition to insurance, organizations must have a well-developed incident response plan in place. This includes understanding their legal obligations regarding evidence preservation, forensic investigations, and the potential involvement of law enforcement. Being prepared for incident response enables those affected to properly handle cyber incidents and minimize the damage caused.

Organizations operating in certain industries must also comply with industry-specific regulations and standards. These regulations, such as the Payment Card Industry Data Security Standard (PCI DSS) or the Health Insurance Portability and Accountability Act (HIPAA), establish cybersecurity requirements that organizations must meet in order to protect sensitive data and maintain regulatory compliance.

Oftentimes, the aftermath of a cyberattack leads to litigation. Affected individuals, shareholders, or regulatory bodies may file lawsuits seeking compensation for damages or alleging negligence on the part of the organization. Understanding liability issues, potential legal claims, and effective defense strategies is crucial for managing legal disputes arising from cyber incidents.

Finally, cyberattacks can originate from anywhere in the world, making matters of international law, jurisdiction, and cooperation with foreign authorities increasingly difficult. Successfully navigating these complexities calls for a thorough understanding of international legal frameworks and solid collaboration with relevant authorities.

To address the multitude of legal issues arising from a cyberattack, cooperation must be fostered between legal teams, IT departments, cybersecurity experts, and external consultants specializing in cyber law and incident response.

Close teamwork ensures a comprehensive and well-informed response to these challenges, enabling organizations to effectively address risks and promptly recover.

The cyberscape presents many constantly evolving pitfalls, demanding a multifaceted approach to security. While robust technical defenses are essential, organizations navigating this terrain must also recognize the crucial role that legal expertise plays in mitigating risk and navigating the aftermath of an attack. In this sense, the legal landscape surrounding cybersecurity is like a minefield.

Understanding the legal frameworks of data breaches, incident response procedures, and potential litigation empowers your leadership to make informed decisions in the face of adversity. Cyber counsel equips your organization with the legal weaponry to defend itself in courtrooms across the globe, should the need arise. These trusted advisors can guide you through whatever legal challenges you face, facilitating a swift and efficient recovery.

The digital battlefield is far from static. With the emergence of advanced threats like AI-powered attacks and quantum computing, legal expertise must be equally dynamic. Collaborating with cyber counsel who continuously update their knowledge and network of connections ensures your organization remains ahead of the curve, able to adapt and anticipate future legal challenges.

Ultimately, cyber counsel is not just a reactive shield against legal fallout; like breach coaches, they are proactive architects of digital fortitude. Their assessments identify vulnerabilities before they are exploited, and their advice strengthens your security posture from within. This holistic approach to cybersecurity, informed by both technical and legal expertise, is the cornerstone of a truly impenetrable digital fortress.

The time to prioritize cyber counsel within your organization is now. Recognizing the legal dimension of the cyberscape and equipping yourself with the necessary expertise is a critical step towards achieving long-term success in the modern age. By fostering a collaborative environment where legal minds synergize with IT teams and cybersecurity experts, you can transform the digital battlefield into a platform for showcasing your organization's resilience and adaptability.

Remember, in the face of cyber threats, failing to prepare is preparing to fail. Invest in cyber counsel, not just for the day the storm hits, but for the enduring peace of mind that comes from knowing your digital borders are patrolled by the best. As you chart your course through the complex legal landscape of cyberattacks, do so with confidence, knowing you have the necessary expertise by your side.

Chapter 12
CYBER INSURANCE

Have you ever considered the potential cost of a cyberattack to your organization? According to recent statistics, cybercrime is projected to cost the world $10.5 trillion annually by 2025. With the increasing reliance on digital assets, it's more important than ever for organizations of all sizes to protect themselves with comprehensive cybersecurity measures.

Yet, no matter how robust your security protocols are, you can never be entirely immune to cyber threats or the negative impact of a breach. Phishing schemes, supply chain compromises, insider threats, ransomware attacks—cybercriminals and nation-state actors have proven remarkably innovative and persistent in identifying and exploiting organizations' vulnerabilities for financial or strategic gain.

In the unfortunate event of a cyber breach, your breach coach and cyber counsel play essential roles in managing the immediate crisis and its aftermath. However, an additional layer of protection comes in the form of insurance. Cyber insurance acts as a financial safety net that helps mitigate the immense cost of recovery, allowing your company to get back on its feet quickly.

When an organization's preventative cyber defenses fall short of blocking intrusions, the fallout can be catastrophic without adequate protection. A breach can trigger massive costs from business interruptions, legal and regulatory actions, ransomware extortion payments, crisis communications, and reputation repair, not to mention liability arising from damages to customers, partners, and other third parties impacted.

Small and mid-sized businesses especially may lack the resources or expertise to effectively respond and recover when disaster strikes, facing possible insolvency if left unprotected. Having comprehensive risk transfer mechanisms in place before an incident occurs is absolutely imperative.

In such a situation, cyber insurance is a powerful tool that provides funds needed for the recovery and continuity of the business, such as ransom, costs for reputational damage repair, regulatory fines, and legal fees resulting from the security breach. It acts as an invaluable financial partner that bears the brunt of these expenses, reducing the burden on the organization.

Think of it as a robust financial shield that deflects legal fees, extortion payments, and regulatory fines. It also covers expenses like data recovery and forensics, regulatory penalties, public relations and crisis communications, and notification costs. These financial implications can often be significantly high, making cyber insurance an ideal investment.

In this chapter, we will undertake a comprehensive examination of cyber insurance, exploring its core protections, how policies are structured, where critical coverage gaps often exist, best practices for integrating insurance into enterprise risk management, and how to ultimately get the most value from your cyber insurance program.

With the right understanding of this continually developing risk transfer tool, you can make well-informed decisions about the appropriate coverage for your organization, as well as implement layered protection tailored to your unique risk profile. By taking a proactive approach to assessing and addressing cyber risk, you can empower your company to continue thriving in today's increasingly treacherous digital ecosystem.

Once considered a niche offering limited to a handful of insurance companies, cyber insurance policies have now become an indispensable part of risk management for businesses of all sizes. They shield against a wide array of cyber threats, such as data breaches, cyber extortion, and business interruption caused by cyberattacks. While the immediate benefit of cyber insurance lies in financial protection, its advantages go beyond that. Its providers are now taking a more proactive role in enhancing their clients' security measures.

The cyber insurance industry is evolving in response to key market trends, including the increased adoption of cloud-based technologies, the rapid expansion of the Internet of Things (IoT), and the mounting frequency of cyberattacks. These trends have given rise to innovative cyber insurance products designed to address emerging risks and protect organizations from the potentially devastating consequences of a cyber incident.

The internet's rise and the escalating number of data breaches and cyberattacks have contributed to a surge in demand for cyber insurance policies. With cyber threats showing no signs of slowing down, a growing number of insurance companies have entered the market and now offer tailored protection against the ever-evolving digital threat landscape.

As the industry expands, major players like AIG, Chubb, Zurich, Allianz, and Berkshire Hathaway have emerged as prominent providers of cyber insurance solutions. These companies offer a diverse array of products and services, including risk assessment, incident response planning, and cybersecurity consulting, catering to the growing demand for comprehensive coverage.

To leverage cyber insurance effectively, organizations must first decode the structure of policies available in the marketplace. These policies contain various components that address different aspects of cyber risk as well as associated financial damages that may result from security incidents. Two principal categories of coverage exist: first-party and third-party.

First-party coverage protects your organization directly. It includes costs involved with investigating and remediating breaches, restoring compromised systems and data, replacing hardware, covering income lost due to business interruptions, paying ransoms, and securing crisis management and public relations services. First-party coverage is essential for managing damages and disruptions to your own systems and data.

Third-party coverage is designed to address damages that your company may cause to others, for which it may be held liable. This can encompass legal defenses, settlements, regulatory fines, and notification costs related to privacy breaches. Given the increasing tendency of cyber threats to spread through supply chains, customer networks, and business partnerships, liability risks have grown exponentially. Collateral damage resulting from cyber incidents can extend far beyond an organization's own boundaries, impacting partners and customers alike. Coverage for third-party liabilities is thus critical to minimize any financial and reputational consequences arising from this spread.

Aside from damages caused by cyberattacks, comprehensive policies also cover preventative services. Cybersecurity risk assessments, vulnerability testing, security training, and threat intelligence resources that help strengthen defenses ahead of time are all components of a well-rounded policy. By tailoring the coverage to fit your organization's individual risk profile, you can strike a balance between reactive and proactive measures, ensuring that you are not only protected during an incident but also equipped to mitigate potential threats before they strike.

Premiums and exclusions are other vital considerations when evaluating policies. They serve as determinants for the coverage. Cyber insurance policies often contain exclusions like incident types, impacts and causes. These on their own do not trigger the coverage. Some common exclusions involve unauthorized access without proven data or system damage, on-premises hazards and physical perils, war or terrorism, and infrastructure outages unrelated to cyberattacks. While some exclusions are standard, insurers may incorporate restrictive ones that decrease a policy's value, so be sure to scrutinize the fine print.

Insurers also conduct rigorous underwriting processes analyzing applicants' cyber risk profiles to inform pricing. Companies exhibiting sound cyber hygiene and data protection pay significantly lower premiums, while higher-risk organizations and industries face more expensive coverage, reflective of their elevated exposure. Reviewing proposed policies in detail is essential to avoid coverage gaps.

To successfully navigate the complex world of cyber insurance, your IT/security team must collaborate with brokers in order to conduct necessary risk assessments. This helps ensure that underwriting is transparent and results in the most favorable terms for your organization. It is also important to take a proactive approach by strengthening your defenses and governance before policy renewals. Doing so can earn you more favorable premiums over time.

Managing cyber insurance is not a task that can be handled by just one department. It requires coordination between various teams such as risk managers, legal/compliance teams, IT/security leaders, crisis response units, and executive leadership. Working closely with brokers helps you negotiate policies specifically tailored to your organization's unique needs and vulnerabilities.

Regular annual policy reviews are needed to keep up with thc ever-changing risks in the cyber landscape. It's important to benchmark your coverage options against others in the market and model different threat scenarios to ensure that your protection stays up to date. While no cyber insurance policy is designed to fit all needs, taking a methodical approach to evaluation and customization allows organizations in all sectors to secure robust, cost-effective policies that align with their risk appetites.

A common misconception is that cyber insurance policies are only useful after incidents occur. However, forward-thinking insurance companies are now adopting a more proactive stance, working closely with clients to fortify their cyber defenses. These innovative insurers offer advisory services and incentives, fostering a collaborative environment that bolsters clients' risk postures.

Many leading insurers are incentivizing policyholders to implement foundational cybersecurity best practices, such as establishing cyber incident response plans, adopting multi-factor authentication, implementing network segmentation, and conducting staff security awareness training. Some insurers have even made these controls mandatory for coverage, promoting a culture of strong cyber hygiene and resilience.

The symbiotic relationship between insurers and insured extends beyond financial protection. Insurers, with their wealth of experience in cybersecurity, offer invaluable perspectives on bolstering clients' vulnerability assessments, penetration testing, compliance audits, and risk quantification efforts. Additionally, the sharing of threat intelligence enhances prevention and early detection capabilities, further solidifying the collective defense against cyber threats.

To further enhance preparedness, leading carriers facilitate breach response exercises that simulate real-world scenarios, evaluate clients' plans and procedures, and provide access to trusted external response teams. This close collaboration between clients and insurers fosters institutionalized muscle memory and unified command structures, enabling a swift and effective response to security breaches.

As technology continues to evolve, the possibilities for enhancing cyber resilience through insurance are expanding to unprecedented levels. Emerging options such as blockchain-based data platforms have the potential to facilitate real-time risk assessment and automated claims processing. Moreover, advancements in data analytics and artificial intelligence may soon allow the prediction and prevention of attacks based on early indicators. As technology forges ahead, the partnership between insurers and policyholders will only grow stronger.

With cyber insurance now prevalent in the market, choosing the right carrier is of utmost importance to maximize value in today's landscape. Numerous factors must be considered when selecting a provider and coverage that aligns with your organization's unique needs.

Cyber insurance serves as your emergency response unit during critical incidents, so while cost remains an important aspect, the emphasis should be placed on the insurers' service capabilities. Proven expertise and responsiveness are the cornerstones of an effective partnership. When evaluating potential insurers, consider their track record in managing major breaches, paying attention to the speed of claims processing and feedback from past clients. This can shed light on their efficiency and reliability in times of crisis.

Specialized expertise in the realm of cyber insurance is another crucial factor to consider. Opt for carriers that boast dedicated cyber practices and underwriters who specialize in this niche field, rather than settling for generalist insurers who merely dabble in cyber coverage.

Effective communication and collaboration are paramount in the insurer-insured partnership. Assess the insurers' client engagement models and seek responsive contacts who dedicate time to understanding your organization's particular risks. Avoid carriers that offer opaque and impersonal interactions, as these may hinder teamwork in an emergency situation.

Financial strength is also key, since this affects whether the insurer remains solvent and capable of paying out major claims. Seek out carriers that are backed by robust reinsurance and hold high ratings from reputable firms such as AM Best.

Lastly, tap into the wisdom of advisors and industry peers when evaluating potential insurance providers. Leverage the insights of consultants who can compare and contrast offerings across different insurers, ensuring that your organization selects the best possible partner to navigate the ever-evolving cybersecurity landscape. By carefully considering these factors and establishing a strong partnership with the right cyber insurance carrier, organizations can fortify their defenses and build a resilient foundation to withstand the challenges of the digital age.

Once prospective partners have been identified, it is imperative to meticulously analyze policy wordings before making a purchase. Scrutinize key factors such as total coverage limits, sub-limits for specific services, deductibles and retentions, exclusions, and conditions that trigger coverage. This analysis enables a trade-off evaluation between premium costs and policy protections. It's important to strike a balance between broader coverage and higher premium investments that align with your budget and risk tolerance.

Remember, selecting the right carrier partner is a pivotal decision that requires careful research. By considering these factors and conducting a thorough analysis, you can make an informed choice that maximizes the value of your cyber insurance coverage.

The moment of truth for any cyber insurance purchase comes when an actual claim is filed. Make sure your organization is prepared to quickly activate policies and effectively collaborate with carriers when breaches occur.

Promptly notifying insurers of a potential claim is critical, as delayed reporting can complicate the claims process. Make sure you know the required timeframe outlined in your policy—usually 48–72 hours—and the proper channels for reaching out. Filing a formal claim should include detailed incident information and follow the reporting mechanisms established in your policy. Maintaining clarity on the claims procedure and timeline helps set expectations.

Be ready to invest time in compiling forensic evidence, documenting losses, and quantifying damages for claim submission. Insurers will thoroughly assess whether the incident meets policy triggers to qualify for a payout. Securing assistance from legal counsel and public relations firms is also prudent for navigating inquiries amid a crisis.

Clear and proactive communications are vital throughout the claims process. Ask questions and provide requested documentation expediently to facilitate processing. While insurers will handle much of the legwork in recovering from an attack, remaining engaged with them will streamline claim resolution.

It's important to understand that while cyber insurance can be a valuable tool in protecting your organization, it should not be relied upon as the sole solution for security. Instead, it should be integrated as part of a comprehensive risk management strategy that includes prevention, detection, response planning, and transfer. To truly strengthen your cybersecurity defenses, foster a culture of awareness and responsibility throughout your entire enterprise. Encourage employees from all departments to practice safe data practices, exercise caution, and quickly report any concerns they may have. Providing extensive training on common threats like phishing, ransomware, and social engineering scams is crucial for frontline defense.

Regularly reassess your insurance coverage to ensure it aligns with your evolving risk profile. As your business ecosystem changes—through new acquisitions or emerging risk exposures, for example—you may need to make adjustments to your coverage. Major technology migrations, like adopting cloud services or integrating operational technology (OT) infrastructure, can also create gaps in coverage. Operational technology refers to the hardware and software systems that monitor and control physical processes, devices, and infrastructure in various industries, such as manufacturing, utilities, and transportation. Given advancements like these, it's important to review your policies regularly against offerings in the marketplace to verify adequate protection for your organization.

Maintaining open and productive dialogues with your insurance carriers is also key. Discuss your risk management initiatives and participate in benchmark assessments to ensure that your policies are priced appropriately based on your organization's changing risk posture. By committing to continuous improvement and engagement with your carriers, you can build lasting trust and maximize the value of your insurance policies over the long term.

Cyber insurance serves as a vital safety net and risk mitigation tool for organizations, offering financial protection and support in the face of cyber incidents and breaches. It helps cover the costs associated with incident response, forensic investigations, legal fees, regulatory fines, reputation management, and even potential lawsuits arising from data breaches. Without cyber insurance, organizations may find themselves struggling to bear the significant financial burden of recovery on their own, making it an essential component of a comprehensive risk management strategy.

As technology rapidly advances, the cyber threat landscape continues to present new challenges. Organizations must stay ahead of them by adapting and revising their insurance policies. This entails regularly reassessing cyber risks, comprehending the coverage provided by their policies, and collaborating with their insurance providers to ensure that the policies align with their evolving cybersecurity needs. By being engaged and watchful, companies can strengthen their resilience against new and emerging threats.

Take proactive measures in evaluating your organization's cyber risks and safeguarding their digital future. Begin by conducting a thorough assessment of your cybersecurity measures, identifying potential vulnerabilities, and implementing robust safeguards. Concurrently, explore the various cyber insurance options available in the market. Engage with reputable insurance providers to understand their policies and how they can cater to your unique requirements.

By combining comprehensive cybersecurity measures with appropriate cyber insurance coverage, you can significantly enhance your organization's ability to withstand and recover from cyber breaches. This approach ensures that your organization is well-prepared to face the ever-evolving cyber threat landscape and minimize the potential impact of security incidents and breaches on your business operations and reputation.

Remember, cybersecurity is an ongoing journey that necessitates continuous efforts to adapt and protect against evolving threats. It is always better to be prepared and secure your digital assets.

Chapter 13
CRISIS COMMUNICATION

Dealing with the aftermath of a cyberattack can feel like you're lost at sea, plunging into uncharted and treacherous waters. Suddenly, your organization is facing choppy waves of operational disruption, financial damages, regulatory scrutiny and public criticism. It's easy to feel overwhelmed. So how do you keep your head above water when navigating a cyber crisis?

One of the most critical aspects of mitigating damage and restoring trust following a cyber incident or breach is effective crisis communication. Your partnership with your chosen cyber insurance company, along with the support of a cyber coach, is vital in navigating a crisis. While the insurance company provides financial and technical support, the cyber coach offers expert guidance on communication strategies to effectively manage the fallout.

However, the level of transparency and communication with all stakeholders, including employees, customers, partners, and the public, will greatly impact the overall outcome. Clear, timely, and empathetic communication is necessary for maintaining trust, minimizing reputational damage, and showing your organization's commitment to addressing the incident responsibly and professionally.

As the saying goes, "Communication is key." This is truer than ever when dealing with cyber crises. Organizations must be well-prepared to communicate with internal stakeholders, customers, business partners, regulatory bodies and the public in a prompt, transparent manner. Failure to do so can lead to further reputational harm, loss of customer confidence in the business, and even legal ramifications.

Having a well-crafted crisis communication strategy is essential, but it is so often overlooked. The way you communicate after an incident plays a huge role in determining how much damage is done to your organization's reputation. It's not just about dealing with the technical incident response—it's also about managing people's perceptions.

To navigate these treacherous waters, it's crucial to embrace key principles that will guide your communication strategy in adversity. They will ensure a swift, empathetic, accurate and well-coordinated response to your stakeholders.

One principle to consider is speed and transparency. During a cyberattack, time is of the essence. Acting swiftly to disclose the incident and its impact demonstrates a commitment to honesty and shows that you have the situation under control. Even if the details are incomplete, sharing updates in real time can help build trust.

Delaying the release of information can foster doubt and speculation, making it harder for you to regain control of the narrative. Therefore, it's vital to take a proactive approach, allowing you to provide accurate and reliable updates that better empower stakeholders to protect themselves.

Consider the example of a financial institution that suffers a data breach. The moment they become aware of the incident, they notify their customers via email and on their website, outlining the steps they are taking to investigate and rectify the situation. This early transparency builds trust and reassures the customers that their interests are being prioritized.

When a breach occurs, it is essential to recognize and acknowledge the concerns of individuals affected by it. Leaders of the organization that is breached must express compassion and signify a commitment to aid stakeholders through these trying times. When you display a genuine understanding of their distress and offer support, they will feel heard and cared for. Showing empathy, rather than addressing purely technical aspects of the incident, fosters a sense of care and understanding.

By incorporating emotional support resources and clear instructions on how to protect personal data, you demonstrate a trustworthy concern for customers' well-being. Consider a hypothetical scenario wherein the leadership of a healthcare organization neglects to incorporate empathy and compassion into its communication strategy. In this fictitious case, the organization experiences a data breach that compromises patients' sensitive medical information.

Rather than recognizing the emotional ramifications of the breach on affected patients, the leadership concentrates solely on technical outcomes. They provide instructions on safeguarding personal data, but fail to exhibit understanding or acknowledgment of the anxiety and stress that the patients may be enduring due to the privacy violation.

The absence of empathy causes the leaders to be perceived as distant and indifferent. Their communication strategy lacks the human touch that proves requisite in such distressing situations. Consequently, the leaders miss an opportunity to connect with affected individuals, leaving them feeling unheard and unsupported.

This lack of empathy can significantly impact patient care and recovery within the hospital. Without the reassurance and support that accompanies empathetic and compassionate communication, patients may feel neglected and even seek care elsewhere, thereby compromising the hospital's patient retention rate. Furthermore, they may be less inclined to disclose critical information or concerns about their health, hindering accurate diagnoses and treatment.

Similarly, the emotional well-being of stakeholders, customers, and the public plays a pivotal role in the recovery process during a crisis. Just as patients may experience heightened stress and anxiety without empathy, potentially delaying their healing and impeding their overall recovery, stakeholders and customers require empathy to cope with the aftermath of a cyber incident. Without the care and understanding provided by empathetic communication, they may struggle to navigate the challenges they face.

The absence of empathy and compassion can also significantly impact the employees within the organization.

When employees witness this absence from their leaders, they may feel disheartened and less motivated to provide the highest quality of care. This, in turn, can negatively affect their job satisfaction and the overall quality of customer care in the organization.

In adverse times, leadership must recognize the importance of empathy and compassion in supporting their staff and incorporate these elements into their communication efforts. By doing so, they can provide the reassurance, help, and understanding needed to facilitate their recovery process.

Accuracy in communication is another important principle for any organization facing a cyber breach. During such an incident, the potential for misinformation and speculation is high, and this can be a major issue. With so much information being shared online, it can be difficult to discern what's accurate and what is not. Unfortunately, this can lead to panic and confusion, which can further exacerbate the crisis.

To combat the spread of rumors and false information, organizations must refrain from contributing to them. The key is to stick to verified facts when sharing news about the cyber breach. This means conducting a thorough investigation and confirming the accuracy of the details before disseminating them to stakeholders. Providing inaccurate or misleading information will only erode trust further.

Consistency in messaging is crucial as well. Organizations should strive to maintain a consistent narrative across all communication channels, such that the information shared aligns with the facts and avoids creating confusion. This strategy creates transparency and credibility, reassuring stakeholders that the organization is united and actively addressing the breach.

Timing and frequency of communication also play a role in ensuring accuracy and consistency. Regular updates, as long as they are well-verified, help manage the flow of information and reduce uncertainty. However, it is important to strike a balance between supplying prompt updates and taking the necessary time to gather accurate data.

When you prioritize accuracy and consistency, your organization establishes itself as a credible source and can effectively mitigate the spread of misinformation, address concerns, and maintain trust with stakeholders during a cyber breach. Close collaboration between cybersecurity teams and PR departments is also essential to thoroughly vet all messages and make sure they align with the organization's goals and values.

Effective crisis communication is an endeavor that extends beyond isolated efforts. A successful response requires coordination among various groups within an organization. IT teams, legal counsel, public relations, executive leadership, and the cyber coach should move forward as one, aligning on messaging and actions to present a unified front during and after an attack. The cyber coach, as an expert advisor, can guide the organization through the upheaval by providing strategic advice, sharing best practices, and facilitating coordination between different teams. Fostering open communication among these diverse teams keeps everyone on the same page, projecting a sense of control that is crucial in times of uncertainty.

Leveraging the expertise of multiple stakeholders ensures a comprehensive and well-coordinated response to cyber incidents. The IT teams provide technical insights and solutions; legal counsel navigates the regulatory landscape; public relations manages external communications; executive leadership contributes strategic alignment; and the cyber coach offers expert guidance to optimize the organization's game plan.

By weaving these perspectives together, your organization can craft a comprehensive communication strategy that addresses the multifaceted aspects of a crisis. Synergy fosters adaptability, allowing your company to respond dynamically to evolving challenges. It not only enhances the effectiveness of communication efforts but also minimizes confusion and strengthens the overall result.

These foundational principles enable organizations to build trust and understanding even amidst chaos. Acknowledging the human impact of an attack and assisting victims builds goodwill. Anchoring communication in facts and consistency prevents confusing mixed messages. Prompt transparency counters the rumors that thrive on uncertainty during crises. Adherence to these principles is vital for navigating chaotic events.

While it may sound simple enough, it takes preparation and practice to get this right in the heat of the moment. To this end, you must conduct simulations and train your key spokespeople in media communications. Have pre-approved holding statements ready to go. Establish coordination across your internal teams and external partners.

Do this crisis planning upfront, and you'll have a life raft to cling to when turbulent waters strike. Mishandle communications, and it may sink customer confidence and your company's reputation. With so much on the line, why leave anything to chance? Steady leadership and human-focused communications will help keep your organization afloat during the storm.

So how exactly should you go about developing this robust crisis communications plan tailored to your organization? First, clearly identify your key audiences and preferred communication channels. Customers may be best reached via email and social media, while media briefings and press releases are appropriate for journalists. Regulators may require formal documentation. Know how each group needs to be engaged.

Next, prepare pre-approved messaging like holding statements, FAQs, and social media posts that can be launched immediately during an incident. Draft templates for crisis emails to internal and external stakeholders as well. Having initial announcements ready to go can help save valuable time.

Next, carefully designate and train media spokespeople who will interface with each audience group. Subject matter experts make ideal spokespersons when provided with communications coaching. You should also establish a crisis command center with cross-functional teams that are empowered to make rapid decisions.

Conduct simulations via tabletop exercises, modeling cyber incident scenarios to practice activating crisis protocols. Run through mock press conferences and media interviews. Regular training ensures your response plan is solid and personnel have confidence executing it during precarious events. With the right planning, your organization can navigate hardships with poise.

Once your crisis plan is in place, effectively managing communications across multiple channels will be critical during an actual incident. For press releases and media briefings, focus on providing journalists with timely, accurate information through proactive statements and press conferences as events unfold. Keep your website updated with the latest factual developments, helpful resources, and progress reports on response efforts.

Social media platforms will also become paramount for sharing important updates, addressing frequently asked questions, and directing stakeholders to additional resources. Since social media offers real-time communication capabilities, you should diligently monitor your accounts and respond quickly to customer concerns with guidance.

For internal communication, give employees first access to updates and encourage feedback. Effective communication with employees is key. It is also important to keep them in the loop on priorities like safety and business continuity. Regular updates through email, internal messaging systems and intranets contribute to transparency about the incident's impact on the organization.

Disseminating necessary precautions and changes in policies or procedures and encouraging open dialogue can foster a sense of collaboration among employees. Establishing channels for employees to ask questions or express concerns is equally important to further strengthen internal communication in the organization.

In customer communication, a proactive approach is vital, as is explaining the event's impact on customers and addressing their concerns without delay. When you utilize various channels, such as emails, customer portals, and dedicated helplines, you ensure regular updates on the crisis resolution's progress. By offering support and resources, you also demonstrate a commitment to customer well-being.

Coordinating messaging across these channels will help to maintain consistency and transparency. You must regularly assess the communication plan's effectiveness and make adjustments as needed. This helps meet evolving stakeholder needs. With a well-executed plan, information flows seamlessly across all channels, and you can navigate through your predicament with confidence.

When managing crisis communications during cyberattacks, there are several common pitfalls you must avoid. One key mistake is delaying disclosure or attempting to conceal the breach, as this can erode your credibility when the truth inevitably surfaces. Even if all details are not known, you should promptly notify stakeholders. Early notification is crucial for preserving trust during an emergency, as it shows transparency and accountability. By demonstrating a commitment to open communication, organizations can help preserve their reputations and build confidence among their stakeholders during challenging times.

Equally critical is providing accurate information. Misreporting facts or speculating without confirmation can lead to embarrassing corrections. Verifying information before sharing it helps prevent the spread of false or misleading details, which also safeguards the organization's reputation.

Sensitivity and empathy towards those affected by the attack must be prioritized to avoid being perceived as indifferent. Acknowledging the impact and seeking to genuinely understand concerns through open listening and meaningful dialogue can help mitigate negative outcomes.

Coordinating communication across channels and spokespeople is important for consistency and to avoid contradictory messaging. Inconsistent messages breed mistrust and confusion. It is therefore vital to maintain unified communication across all channels and aligned responses from spokespeople. This reinforces the organization's credibility and encourages trust.

While social media is a significant tool for crisis communications, it can contribute to the rapid spread of misinformation. Organizations must vigilantly monitor social platforms and swiftly address inaccuracies. Social media can also be a platform for stakeholder outrage and protests, which necessitates careful engagement to avoid inflaming negative sentiments. There is a need for prudent policies and training guides for spokespersons navigating social media during situations like these.

Despite its potential pitfalls, social media remains invaluable for providing real-time updates and reaching a wide audience during disasters in the digital age. By maximizing its reach while minimizing associated risks, organizations can effectively manage and withstand the aftermath of an attack.

Shifting the focus to the globalized digital landscape, organizations must pay attention to cultural differences and language barriers during a crisis. Cultural intelligence is crucial, and diverse communication styles and norms must be considered. Adapting messaging to align with cultural sensitivities helps ensure that all audiences feel understood and respected.

Language barriers can pose challenges, and providing translations or multilingual support is key to overcoming this hurdle. Investing in sensitive and accurate translation is vital to avoid misinterpretations that could worsen problems or cause unnecessary confusion.

Successfully navigating these additional considerations of a global internet requires both organizational and interpersonal skills. Cultural competency, a deep understanding of target audiences, and proactive management of social media dynamics are essential for effective crisis communications strategies. By incorporating these components, organizations can enhance their response and ably address the difficulties posed by cyberattacks.

We can learn valuable lessons on effective crisis communication by examining real-world examples. Take the Equifax breach, for instance. Despite the massive impact on millions of Americans, the credit reporting agency took the initiative with the CEO publicly apologizing, offering free credit monitoring, and outlining plans for improved cybersecurity. This conveyed accountability and a commitment to rebuilding trust.

In the case of Facebook's Cambridge Analytica scandal, Facebook took decisive steps to address the situation. The company implemented significant reforms, such as centralizing privacy controls and providing comprehensive transparency reporting. These measures were aimed at restoring confidence in Facebook's promise of user privacy.

Maersk, a global shipping company, dealt with a devastating ransomware attack that paralyzed its operations worldwide. However, the company's leaders openly shared accounts of the challenges they faced and the resilience required to restore critical systems within just 10 days. By sharing their experiences, Maersk demonstrated their commitment to overcoming the attack and served as a source of inspiration.

Conversely, Uber's 2016 data breach serves as a cautionary tale due to its mishandling. The company concealed the breach for a year and even paid hush money to the hackers involved. When the breach finally became public, Uber's apologies were met with skepticism and did little to alleviate the damage caused by their initial lack of transparency. This case highlights the consequences of delayed disclosure and deceptive communication during a cyber incident.

These examples emphasize the importance of responsible disclosure, decisive action, and empathy in crisis communication. Promptly and honestly addressing the situation, taking concrete steps to mitigate the impact, and demonstrating empathy towards the affected parties can help preserve trust and credibility.

On the other hand, delaying or obfuscating communication only serves to erode goodwill and exacerbate problems further. By learning from these examples, organizations can enhance their crisis communication strategies and handle cyberattacks while maintaining the trust of their stakeholders.

When trouble emerges, effective communication is the lifeline that can help your organization carry on. By dedicating time and resources to develop crisis readiness and response plans, you can fortify your company's foundation, enabling it to weather any unexpected challenges. Much like constructing a sturdy building that can withstand natural disasters, investing in crisis preparedness ensures your organization remains stable and strong when adversity strikes. This proactive approach not only protects your company's reputation but also fosters trust among stakeholders, employees, and customers.

Effective communication is ultimately about preserving relationships under the most difficult circumstances. With preparation and care, organizations can retain composure and integrity during times of hardship.

Key takeaways include adhering to core principles of transparency, accuracy, empathy and consistency in all messaging. Develop robust crisis plans detailing appropriate channels, spokespersons and operational response for all stakeholders. Train personnel extensively via simulations and media coaching. Actively manage communications across owned, shared and earned media channels, while avoiding issues like delays, misinformation and internal fragmentation. Focus on steady leadership and compassionate correspondence.

By instilling crisis readiness into organizational culture, engaging stakeholders honestly, and dedicating resources for the long haul, companies reinforce their resilience. With the right planning and people, you can navigate the stormy seas and come out stronger. When it comes to cyber crises, preparation and communication are everything.

Now that we have covered essential security measures to take during and after a cyberattack, it's time to look ahead. In the coming chapters, we will explore critical controls and strategies to address the growing threat of ransomware. We'll also examine the future of cybersecurity, emerging technologies, and how organizations can stay adaptable in ever-evolving digital landscapes.

While the cyber terrain may seem rough at times, with vigilance, care and collaboration, we can chart a course through the risks and opportunities ahead.

Chapter 14

THE ESSENTIAL CONTROLS REQUIRED TO PREVENT RANSOMWARE

Ransomware threats have reached unprecedented levels of disruption. As discussed in the previous chapters, these attacks have evolved to utilize sophisticated techniques that circumvent conventional security measures. Faced with this escalating threat landscape, organizations must adopt a cybersecurity approach that goes beyond reactive defenses.

In this chapter, we will explore the controls and strategies needed to bolster ransomware resilience. Understanding these proactive safeguards and implementing them as a unified defense is crucial for security in today's digital era.

The rising pace and impact of ransomware attacks make their prevention an urgent priority. Beyond financial losses, the fallout extends to reputational damage, legal liabilities, and loss of customer trust. To mitigate their multifaceted consequences, a layered cybersecurity strategy is required, addressing vulnerabilities across data protection, access controls, email defenses, and network security.

The controls outlined in this chapter are selected for their proven effectiveness against ransomware campaigns. While no single control offers total protection, collectively they reduce risk exposure and minimize potential damage. It is imperative to recognize that ransomware prevention requires a comprehensive approach rather than reliance on individual controls in isolation.

In addition to technical controls, the human element plays a key role. Employee cybersecurity awareness and training are vital for creating a culture resistant to the social engineering that often facilitates ransomware attacks. Investing in education and building staff awareness fortifies a company's human defenses.

In the sections that follow, we will delve into each essential control, providing practical guidance for implementation. By examining their interdependencies and synergies, organizations can construct an integrated defense tailored to combat ransomware's continuous evolution.

Ransomware resilience is imperative to protect businesses from crippling operational disruption. Let us embark on building the knowledge and cybersecurity foundations needed to defend against this threat in an ever-changing cyber landscape. With preparation and vigilance, we can counter this menace and ensure our digital safety.

Data Backup and Recovery

Comprehensive data backup and recovery (BCDR) strategies are a fundamental defense against ransomware. Regularly backing up critical data and having reliable recovery plans in place allows organizations to mitigate the impact of attacks.

In the event of a ransomware incident, dependable backups enable quick restoration of systems without paying ransom demands. It provides an alternative to capitulating to extortion. A successful backup strategy requires identifying high-priority data and determining the optimal backup frequency. Frequent backups ensure access to current data versions when needed during an emergency.

Secure offsite storage of backup data limits its exposure to the same risks as primary data. Cloud storage or external media offers isolation from the organization's main infrastructure. Recovery plans outline detailed procedures for restoring data, roles and responsibilities, and processes for stakeholder notification. Regular testing helps verify the plan's effectiveness.

In essence, reliable data backups serve as a cybersecurity safety net, allowing organizations to confidently recover from ransomware attacks. They protect against digital extortion, just as physical safes protect valuables. This was demonstrated in real-world incidents like the cyberattack on Baltimore, where the city refused ransom demands due to well-planned backups. Despite high restoration costs, the ability to recover from backups created resilience.

Endpoint Detection and Response (EDR)

While data backups provide post-attack recovery, endpoint detection and response (EDR) offers real-time prevention capabilities directly on endpoints. EDR is another critical control for fortifying defenses before damage occurs.

EDR utilizes advanced analytics and machine learning to identify suspicious endpoint activities that could indicate ransomware threats. By flagging potential incidents early, organizations can respond swiftly to prevent wide-scale encryption of systems. EDR also aids investigations by providing visibility into abnormalities and tracing the root causes of issues. Automated response mechanisms allow EDR solutions to isolate threats and prevent propagation to other endpoints.

However, EDR should function as part of a larger security ecosystem including high-quality antivirus, patch management, and user training. Layering it with other controls amplifies its preventative and investigative potential.

Network Segmentation and Access Control

Cybercriminals often try to exploit vulnerabilities in an organization's network infrastructure. Network segmentation and access control help safeguard sensitive data and systems by limiting unauthorized access.

Effective implementation starts with evaluating the network architecture and defining logical boundaries or zones based on data sensitivity or organizational functions. Isolating these zones via firewalls, VLANs (virtual local area networks), subnets, or other controls restricts communication between them. VLANs are a network segmentation technique that logically group devices on the same physical network, allowing them to communicate as if they were on separate networks. This separation adds an extra layer of security and helps manage traffic.

Access control mechanisms then become the gatekeepers, managing and controlling information flow between segmented zones. Access control lists (ACLs) define which users or systems can access specific resources within the network. Robust authentication through multi-factor and role-based access control adds additional layers of security for network access. These mechanisms reduce the risk of unauthorized entry or lateral movement across the network.

Maintaining vigilance through regular permission reviews, unnecessary access removal, and preserving segmentation boundaries is key to sustaining effectiveness over time. Additionally, intrusion detection and prevention systems (IDS/IPS) reinforce segmentation by actively monitoring traffic, detecting suspicious activities, and automatically preventing potential breaches. Regular penetration testing and vulnerability assessments identify and address gaps before incidents occur. Simulating real-world attacks evaluates the network's resistance level.

Patch and Update Management

While network controls focus on external threats, patch and update management addresses risks from within by fixing internal system vulnerabilities. It involves systematically identifying, prioritizing, testing and deploying relevant patches or updates for operating systems, applications, and hardware components. Regular scanning and assessments pinpoint vulnerabilities that require patching by comparing configurations against known exploit databases. Prioritization focuses efforts on addressing high-risk flaws first.

Rigorous testing in non-production environments preemptively smooths out any incompatibilities before patch deployment. This minimizes disruptions to operations and end users. Once testing concludes, patches are rolled out in a controlled and coordinated manner with stakeholder approvals. Tracking and verification post-deployment confirms successful patch application.

Continual monitoring of advisories and notifications allows new patches to be evaluated and deployed promptly as needed. Staying up-to-date bolsters ransomware resilience.

Incident Response Planning

Incident response planning serves as a vital ransomware prevention control by establishing protocols to mitigate attack impact. Although covered in-depth previously, its significance bears reiterating. Effective plans identify potential threats upfront and define roles, responsibilities, and systematic response procedures. This includes containment measures to prevent ransomware spread, investigation steps to determine scope and origin, and recovery processes to restore affected systems and data.

Communication protocols for timely internal team, customer, and authority notification are built into the response plan. This ensures coordination and transparency in case of an attack. By outlining the end-to-end incident handling process in advance, response plans minimize disruptions caused by ransomware attacks through swift and systematic execution.

In the broader context of ransomware defense, planning how to respond to potential incidents proactively limits damage. It complements preventative controls by preparing containment and eradication measures in advance, much like a well-prepared emergency evacuation plan for a building. This ensures that when a crisis occurs, everyone knows their role, where to go, and how to best respond. Pairing incident response plans with awareness training strengthens preparation by teaching employees to recognize and report threats early. Empowering staff reinforces human-based cyber defenses.

Application Whitelisting

While response plans mitigate downstream impact, application whitelisting aims to prevent attacks earlier in the cyber kill chain by restricting unauthorized software execution.

Whitelisting allows only approved applications to run on organizational systems, blocking potentially malicious software like ransomware. This limits attack surfaces by permitting trusted apps and denying everything else. By maintaining a comprehensive inventory of authorized software and vigilant monitoring to detect whitelist violations, organizations can significantly reduce their exposure to threats.

Effective implementation of application whitelisting requires regular updates to the list as new legitimate applications arise. In essence, whitelisting flips the script on adversaries by establishing positive security models that permit only known good applications instead of blocking known bad ones. This precautionary approach provides particular protection against zero-day ransomware strains.

Zero-day ransomware exploits previously unknown vulnerabilities, leaving organizations with little or no time to develop and deploy patches or other defenses.

Combining whitelisting with other controls, like patch management and network segmentation, adds layered defenses that significantly enhance ransomware resilience through redundancy. It epitomizes the cybersecurity adage that prevention is ideal, but detection is mandatory.

Secure Coding Practices in Software Development Lifecycles

Ransomware prevention extends beyond operational controls to encompass secure software development processes. While controls like whitelisting aim to prevent ransomware execution, integrating security practices into internal software development can mitigate vulnerabilities that permit attacks.

Building security intrinsically into applications provides an extra layer of defense. It starts with educating developers on coding techniques like input validation, proper error handling, and secure authentication mechanisms. This lays the foundation for writing attack-resistant code.

Performing regular code reviews, static analysis, and vulnerability testing during the software development lifecycle enables proactive identification and remediation of flaws before release. Post-deployment, prompt patching, upgrades, and assessments are necessary for addressing newly discovered vulnerabilities and maintaining the integrity of software. This prevents adversaries from capitalizing on unpatched bugs.

In essence, prioritizing security throughout the entire development lifecycle produces inherently more secure software. Addressing vulnerabilities ahead of time reduces reliance on downstream controls to protect flawed applications.

Much like constructing earthquake resistant buildings, organizations should develop applications with inherent ransomware resilience rather than simply attempting to anchor them afterwards. This inside-out approach to security complements traditional outside-in controls.

Supply Chain Threat Management

As ransomware techniques advance, organizations must also expand security perimeters to address vulnerabilities stemming from third-party partnerships. Supply chain threat management helps mitigate these external weak links.

It begins with conducting thorough risk assessments of all vendor relationships, categorizing them based on potential business impact. More scrutiny should be placed on critical suppliers, giving due diligence to their security postures and incident response capabilities.

Continuous monitoring of vendor security practices over the entire lifecycle of the engagement is crucial to identify emerging issues ahead of time. Regular audits, assessments, and reports serve to uphold standards. Developing and enforcing strict security and compliance policies and contracts is key as well. These documents codify protections required by your organization and outline consequences for violations.

In the event of a vendor-related security incident, well-defined response plans facilitate swift containment and communication. They also outline steps for investigating, resolving, and notifying impacted stakeholders.

Third-Party Risk Management

As organizations embrace digital transformation and rely on an expanding network of vendors, third parties can unintentionally introduce ransomware risks if not properly secured. Implementing third-party risk management (TPRM) controls addresses this danger.

Effective TPRM starts with identifying all vendor relationships and categorizing them based on criticality to prioritize security oversight. More scrutiny is focused on suppliers with high potential business impact. Comprehensive due diligence is then conducted evaluating third parties' cybersecurity postures, compliance, and incident response plans. This insight informs risk levels associated with each relationship.

Ongoing monitoring, audits and assessments of vendor security practices throughout the partnership provides long-term security. Regular reporting and reviews ensure that standards are sustained.

Developing and enforcing strict security requirements through policies and contracts is crucial for mandating protections from outside parties. Detailed agreements outline the required measures, communication protocols and liability for non-compliance. By taking a holistic, anticipatory approach, TPRM reduces third-party cyber risks that could lead to ransomware incidents. Diligently assessing and monitoring vendors fortifies external defenses.

Security User Awareness Training

While TPRM focuses on third parties, security user awareness training (SUAT) strengthens resilience by empowering internal users to defend against ransomware.

SUAT trains staff to identify and resist social engineering techniques, like phishing, which enable ransomware intrusions. Having employees who can spot suspicious emails and links significantly improves attack prevention.

Additionally, SUAT keeps users updated on evolving ransomware trends and tactics so they can recognize new threat patterns. Adversaries constantly refine their methods, so ongoing training is key. Most importantly, effective SUAT fosters a cyber-aware culture where employees are actively engaged as part of the incident response team, not just as a source of risk.

Asset Inventory and Visibility

Building on the concept of proactive defense through SUAT, asset inventory and visibility serve as the bedrock for various cybersecurity measures. Organizations need a precise inventory of their hardware, software, network devices, and other digital assets in order to protect themselves.

Asset inventory begins with the identification of all devices and applications within the organization's ecosystem. This includes servers, endpoints, IoT devices, and software applications. IoT refers to the network of physical devices, such as smart appliances or industrial sensors, that are connected to the internet and can collect, send, and receive data. Accurate documentation of these assets ensures that security teams are aware of the digital landscape they are tasked with protecting.

On the other hand, visibility involves more than just listing assets; it requires understanding their interdependencies and communication patterns. This holistic view aids in identifying potential vulnerabilities or weak points in the network, enhancing overall security.

After the initial inventory, organizations must implement robust controls for continuous monitoring of assets. This means tracking changes, updates, and any new additions to the digital infrastructure. Regular audits and assessments guarantee that the inventory remains accurate and up-to-date.

Continuous Monitoring and Detection

As ransomware attacks grow in frequency and impact, rapid threat detection is vital for mitigating damage. Continuous monitoring and detection provides the real-time visibility needed to identify threats early.

This process involves constantly collecting and analyzing security data from networks, endpoints, servers, applications and other systems to uncover anomalies that may represent ransomware activity.

Advanced technologies like intrusion detection systems (IDS), security information and event management (SIEM) and endpoint detection and response (EDR) enable comprehensive 24/7 monitoring across the environment.

EDR and SIEM utilize correlation analysis and machine learning to pinpoint threats within massive data volumes that human analysts cannot process alone. Automated alerts notify response teams of high-priority incidents requiring immediate investigation.

By continuously monitoring data rather than performing periodic scanning or audits, malicious activities are spotted in real-time, before adversaries have time to initiate encryption or other destructive tactics. Early visibility is key for prompt containment.

Security Incident and Event Management (SIEM)

Security incident and event management (SIEM) is a cybersecurity solution that plays a pivotal role in the early detection, rapid response, and in-depth analysis of security events. SIEM combines software and processes that enable organizations to collect, analyze, and act on security-related data from across the digital environment. By aggregating and scrutinizing data from diverse sources such as endpoints, servers, and firewalls, SIEM solutions identify patterns and anomalies that may indicate malicious activity or security incidents. This capability lets organizations promptly and effectively respond to potential threats like ransomware.

Implementing SIEM requires customizing it to your organization's unique environment and integrating it with other security measures. This tailoring minimizes the solution's false positives, focuses it on critical threats, and enhances its accuracy in detecting and responding to incidents. SIEM is a vital component of an organization's cybersecurity strategy, since it operates as a centralized hub for monitoring, analyzing, and addressing potential threats, streamlining the process and improving overall security posture.

Identity and Access Management (IAM)

Identity and access management (IAM) is a cybersecurity framework that focuses on managing and securing digital identities, ensuring that only authorized individuals have access to sensitive systems and data. It encompasses a set of policies, processes, and tools that work together to govern user identities and regulate access to resources within an organization. By implementing IAM, organizations can mitigate the risk of unauthorized access and insider threats.

Key IAM capabilities include provisioning and deprovisioning access, monitoring privileged users, and enforcing multi-factor authentication (MFA). Provisioning and deprovisioning involve granting or revoking user access to systems or resources based on their roles within the organization. Monitoring privileged users helps organizations track those with elevated access so their activities can be supervised and audited. MFA adds a layer of security by requiring users to provide two or more forms of identification before accessing resources.

As an ongoing process that adapts to evolving threats, IAM helps organizations establish robust, secure access management, enhancing their defenses against cyber threats such as ransomware. By adhering to the principle of least privilege, which means granting users only the minimum access necessary for their roles, IAM enables organizations to maintain better control over their digital environment and minimize the risk of unauthorized access or breaches.

Multi-Factor Authentication (MFA)

Multi-factor authentication (MFA) fortifies user access by requiring users to authenticate their identities through multiple factors, rather than relying solely on passwords. By doing so, it integrates additional layers of security and verification, strengthening cybersecurity defenses against ransomware threats.

MFA draws on diverse authentication methods including biometrics, one-time codes, smart cards, and push notifications. This is particularly helpful for securing privileged accounts and remote access points, significantly reducing the risk of unauthorized entry that could enable ransomware attacks. It also aligns seamlessly with the principle of least privilege—only allowing users to possess the minimum access required for their role. Its implementation to protect critical systems and sensitive data access acts as a major deterrent, making it much harder for ransomware to target high-value assets.

In addition to its technical security benefits, MFA promotes greater user awareness and accountability around authentication practices. Users become more active in securing their own accounts, instead of relying on simple passwords alone. This improves the overall organizational security posture.

Integrating MFA is a strategic imperative for organizations aiming to diminish the risk of ransomware threats. By enforcing stronger identity verification through multiple factors, MFA transcends a simple authentication mechanism, evolving into a proactive security approach that shields critical assets.

Privileged Access Management (PAM)

While MFA acts a as vigilant sentinel against unauthorized access, privileged access management (PAM) provides precise control over higher-level users. It focuses on overseeing and controlling privileged accounts—those which wield extensive access to and control over critical systems and data.

A key principle in PAM is the concept of just-in-time elevation, or ensuring that users only receive elevated privileges for the duration needed to perform specific tasks. This approach minimizes the window of opportunity for adversaries to exploit elevated access, reducing the risk associated with prolonged privileged account exposure.

MFA is itself a fundamental component of PAM, providing an extra layer of security for privileged accounts by requiring multiple forms of identification. Continuous monitoring of privileged user activities is another critical aspect of PAM. By scrutinizing actions taken by users with elevated privileges, organizations can swiftly detect any anomalous behavior indicative of a ransomware attack or other malicious activities.

Prompt deprovisioning of privileged access once it is no longer required is another best practice in PAM. This means that accounts with elevated privileges are only active when necessary, reducing the overall attack surface. Regularly reviewing and adjusting privileged access based on job roles and responsibilities contributes to a safer, more efficient access management strategy.

Secure Email Gateway (SEG)

A secure email gateway (SEG) is a cybersecurity solution that safeguards organizations from cyber threats that frequently use email as an attack vector. The SEG serves as a checkpoint for incoming and outgoing emails, carefully filtering and monitoring them to prevent dangerous content from compromising the organization's systems.

Taking proactive steps beyond simple response, SEGs use sophisticated filtering systems to thoroughly examine emails for indications of malicious intent in attachments, links, and content. This approach enables the SEG to identify and block potential threats before they reach end-users, thereby reducing the risk of security breaches.

SEGs embrace advanced technologies such as machine learning to adapt and evolve with the ever-changing landscape of email-based attacks. This continuous learning allows the SEG to stay ahead of emerging threats and new variations of malicious payloads. By doing so, it complements the human element—providing a layer of defense against social engineering tactics through aiding in the detection of phishing attempts, fraudulent emails, and other socially engineered attacks that often precede ransomware incidents.

Incorporating SEGs into the cybersecurity framework not only prevents potential security breaches but also enhances overall email hygiene. It serves as a crucial control for organizations aiming to bolster their defenses against ransomware and other email-borne threats.

Vulnerability Management

Vulnerability management (VM) is a critical cybersecurity process that helps organizations identify, assess, and remediate vulnerabilities in their digital infrastructure. This covers things like operating systems, applications, and other software components. Rather than being a specific piece of equipment or software, VM is a comprehensive approach that uses various tools, techniques, and processes to ensure an organization's assets are protected against potential exploits and threats.

The initial step in VM is to identify potential weaknesses in the organization's digital environment, categorizing them based on severity and potential impact on the organization's security. This process often involves the use of vulnerability scanning tools and other assessment methods to detect and catalog known loopholes.

Rigorous testing is then conducted to validate patches and fixes before deployment, ensuring that vulnerabilities are addressed without compromising critical systems or causing unintended disruptions. This step helps to maintain the stability and security of the organization's digital infrastructure as problems are remediated.

Furthermore, VM requires continuous monitoring of emerging threats and the broader cybersecurity landscape. Regular vulnerability assessments and penetration testing reveal potential flaws as they are discovered, allowing organizations to stay ahead of evolving threats and address new weak points.

In essence, VM serves as a forward-thinking defense mechanism, reducing the attack surface by promptly addressing known vulnerabilities and anticipating potential threats. By incorporating VM into their cybersecurity strategies, organizations enhance their resistance to cyberattacks and demonstrate commitment to staying ahead of the ever-changing threat landscape.

Threat Hunting

Threat hunting is a proactive security process that involves dedicated IT professionals, often called threat hunters, who use specialized tools and techniques to identify signs of compromise or anomalous behaviors within an organization's digital environment. This specialized role requires a combination of technical expertise, a deep understanding of cyber threats, and familiarity with the organization's network and systems.

Threat hunters leverage various tools tailored for threat hunting, including security information and event management (SIEM) systems, endpoint detection and response (EDR) solutions, and specialized threat hunting platforms. These tools enable hunters to collect and analyze vast amounts of data, such as logs, network traffic, and endpoint activity, to identify potential threats that may evade traditional security measures.

For example, threat hunters may use log analysis to spot irregular inbound network connections, which could signal botnet callbacks often associated with ransomware. Behavioral monitoring and machine learning techniques can also be employed to detect high-risk activities, such as unusual user behavior or the execution of unknown binaries, which might indicate the presence of ransomware or other cyber threats.

Threat hunting plays a crucial role in preventing and detecting ransomware attacks, but it also supports incident response and forensic analyses after an attack has occurred. Threat actors often employ anti-forensic techniques to cover their tracks and evade detection. Threat hunters can uncover these patterns and expose the full extent of a breach, enabling organizations to recover more effectively and implement measures to prevent similar attacks in the future.

Cloud Security

In the era of digital transformation, organizations are increasingly adopting cloud technologies to drive innovation and agility. However, these benefits come with new challenges in protecting cloud environments against ransomware threats.

Cloud security encompasses controls and best practices that aim to safeguard data, applications, and infrastructure hosted in the cloud. A robust identity and access management (IAM) strategy forms the foundation. It ensures that only authorized users and devices can access cloud resources through role-based access controls and strong multi-factor authentication (MFA). This prevents unauthorized access, which is a major ransomware risk.

Encrypting data both in transit and at rest provides additional protection. It guarantees that sensitive information remains secure from unauthorized access or theft during an attack. Network segmentation and access controls in cloud environments help contain incidents by limiting ransomware's ability to move laterally. This segmenting of networks and restricting of access prevents the spread of malware.

Cloud-specific security services like secure web gateways, firewalls, and intrusion detection can be added to offer real-time threat detection and continuous monitoring attuned to the cloud's dynamic nature.

Cloud Security Posture Management (CSPM)

Cloud security posture management (CSPM) is an automated, cloud-based solution that continuously monitors and manages the security posture of an organization's cloud infrastructure. CSPM tools run on cloud platforms, integrating with major cloud service providers like Amazon Web Services (AWS), Microsoft Azure, and Google Cloud Platform (GCP) to ensure cloud resources are securely configured and compliant with industry standards and best practices.

CSPM works by leveraging cloud provider APIs to access and assess the configuration of cloud resources, such as virtual machines, storage buckets, and databases. It then compares these configurations against predefined security policies or benchmarks, such as the Center for Internet Security (CIS) benchmarks or the National Institute of Standards and Technology (NIST) guidelines. The CSPM tool identifies misconfigurations and deviations from the established security baselines, which could potentially expose the organization to ransomware attacks or other cyber threats.

Once CSPM identifies misconfigurations, it generates alerts and provides actionable guidance to help IT and security teams remediate the issues promptly. This process includes identifying the specific problem, but also recommending the steps needed to correct the issue and bring the resource back into compliance with the security policy. By automating the process, CSPM enables organizations to address vulnerabilities more quickly and consistently than with manual methods.

In addition to real-time monitoring, CSPM solutions often provide features such as reporting, auditing, and integration with other security tools like security information and event management (SIEM) systems. This helps organizations gain a comprehensive understanding of their cloud security posture and streamlines the process of addressing any identified risks.

Cloud-Native Application Protection Platform (CNAPP)

A cloud-native application protection platform (CNAPP) is an integrated security solution designed to protect cloud-native applications and their underlying infrastructure throughout their lifecycles. As containerized cloud-native applications become more prevalent, ransomware threats have adapted to target their unique characteristics. CNAPPs address these challenges by securing the entire container lifecycle, from building images to deployment orchestration.

CNAPP solutions are typically cloud-based or deployed as on-premises software. They integrate with popular cloud platforms and container orchestration systems such as Kubernetes and Docker, as well as serverless computing environments like AWS Lambda. CNAPP addresses the unique security challenges of cloud-native applications by providing a combination of preventive and detective security controls, including:

Container image scanning, where tools analyze container images for known vulnerabilities, misconfigurations, and embedded secrets. By identifying potential issues, organizations can ensure that only secure and compliant images are deployed to production environments, reducing the risk of compromise.

Runtime protection, which offers real-time monitoring and protection of running containers. This capability enables the detection and blocking of suspicious activities or abnormal behavior that could indicate the presence of ransomware or other cyber threats.

Secure orchestration, which involves the integration of CNAPP with container orchestration platforms like Kubernetes. This allows for the enforcement of security policies and best practices, such as least privilege access, network segmentation, and secure communication between containers.

Serverless security, which tackles the specific challenges of securing serverless functions. CNAPP detects potential threats and maintains the overall security of the cloud-native environment through isolating these functions, monitoring their behavior, and enforcing least privilege access.

By integrating these security controls into DevOps pipelines, CNAPP allows organizations to address security issues early in the development process. This minimizes the attack surfaces exposed to evolving ransomware threats and helps organizations keep their cloud-native environments safe.

As cyber adversaries continue to evolve their techniques, organizations must implement integrated defenses spanning people, processes, and technologies to stay ahead of the threats. CNAPP is a comprehensive security approach tailored to the unique needs of cloud-native applications and their underlying infrastructure, providing specialized security capabilities throughout the life of each application.

In this chapter, we have explored and gained a comprehensive overview of controls to combat the growing threat of ransomware in today's interconnected landscape. While there is no singular control that offers absolute protection, collectively they reduce vulnerabilities and minimize damage when incidents occur.

Regular training ensures users remain vigilant against social engineering, which is the foremost ransomware infection vector. Proactive threat hunting serves to get ahead of emerging threats, and multi-layered technological controls, including whitelisting, microsegmentation, and encryption, provide redundancy.

Most importantly, cybersecurity must be recognized as an adaptive journey rather than a fixed destination. Through continuous awareness, education, and collaboration, organizations can cultivate resilience against ransomware's inevitable changes. Integrating essential controls into a cohesive and consistent cybersecurity program offers the most effective stance against persistent threats.

In closing, defending against ransomware necessitates constant learning and adaptation aligned with the organization's unique risk profile and priorities. By embracing controls proactively and holistically, risks can be pragmatically managed even when they cannot be avoided entirely. With preparation, vigilance, and resilience, organizations can protect their critical assets and confidently navigate the turbulent waters of today's digital world.

Looking ahead, one emerging challenge is securing new technologies against potential cyber threats that seek to exploit vulnerabilities. As innovations like artificial intelligence, IoT, and quantum computing continue to develop, so too do the tactics of malicious actors aiming to abuse them. This highlights why cybersecurity must constantly evolve in parallel with technological progress.

In the next chapter, we will delve into the unique security challenges posed by emerging technologies, as well as the forward-thinking strategies needed to protect critical systems and data against these threats. While newer technologies provide immense opportunities, they also open new attack surfaces that organizations must be prepared to secure.

Chapter 15
EMERGING TECHNOLOGY MEANS EMERGING THREAT

The rapid advancement of technology brings with it not only the promise of a better future but also the potential for new threats. This duality is particularly evident in the realm of cybersecurity, where emerging technologies like self-driving vehicles, artificial intelligence, and 5G networks hold immense potential for positive change across various sectors, while also introducing novel attack surfaces for malicious actors. As we embrace these advancements, we must recognize the race against time to secure them—a challenge that forms a central theme in the broader Cybersecurity Endgame.

One example of new technologies' possible risks is the imaginary scenario of a self-driving semi-truck being hacked during rush hour, causing accidents and injuries. Although this specific incident is fictional, it underscores the real dangers that can arise from the exploitation of vulnerabilities in advanced systems. While there have been instances of connected vehicles being hacked in controlled settings, such as the 2015 Jeep Cherokee incident, it is crucial to recognize that all emerging technologies can be susceptible to similar threats, particularly when not properly safeguarded.

As emerging technologies continue to evolve, so too do the threats they face. This creates a race against time, as malicious actors exploit their vulnerabilities faster than traditional security strategies can adapt. As soon as a new technology gains adoption, cybercriminals are already searching for weaknesses to infiltrate.

As we enter an era of unprecedented innovation, the rush to secure these technologies gains increasingly high stakes. This is particularly important in the context of America's role in the Cyber Race, where the country's leadership in technology and cybersecurity can have a significant impact on the global stage.

While we continue to embrace the promise of emerging technologies, we must remain vigilant of the risks their adoption may present. This vigilance is a key aspect of the Cybersecurity Endgame, as it allows us to create a world where the benefits of cutting-edge technologies are maximized while their risks are effectively mitigated. By taking a proactive approach, we can ensure that security measures keep pace with the rapid progress of innovation, enabling us to harness the full potential of these advancements without compromising safety.

A fundamental challenge in achieving this goal is the tendency of emerging technologies to prioritize innovative functionality over integrated security during their initial design and development stages. In the race to capture market share and user attention, manufacturers often focus on showcasing exciting capabilities and captivating user experiences, leaving robust security measures as an afterthought. This oversight can lead to the release of products with inherent vulnerabilities, creating a technical debt that falls on both consumers and enterprises that adopt these new tools.

For instance, many Internet of Things (IoT) devices have insecure default passwords, lack encryption for data transmission, and lack basic access controls in their early versions. The emphasis is placed on capabilities and ease of use rather than stringent security measures. This neglect during the initial development stages creates openings for threats that could have been avoided if best practices were incorporated into the foundation.

To overcome this issue, manufacturers must prioritize security from the very beginning, ensuring that it evolves alongside new feature development. By incorporating "security by design" principles into their product foundations, they can address potential threats before they become significant. This approach is particularly important in the context of emerging technologies, whose novelty and rapid evolution present unique security challenges.

Due to the new and swiftly changing nature of these technologies, there is often a general lack of expertise in identifying their vulnerabilities and predicting potential exploits. Both their developers and the organizations who implement them may lack full awareness of the potential security risks involved, making it difficult to foresee how criminals and adversaries might weaponize them. With the exponential adoption of these technologies, threats emerge faster than informed defenses can be developed.

For instance, the emergence of deepfakes, powered by artificial intelligence, has introduced unforeseen pitfalls related to manipulated media. Similarly, the expansion of 5G cellular networks has brought about new weaknesses and signaling risks, due to the increased connectivity speed and unfamiliar protocols. These examples illustrate the importance of collaboration, education, and knowledge-sharing among industry experts, academics, and policymakers to create a more informed and proactive approach to securing emerging technologies.

To bridge the knowledge gap and overcome these hurdles, we need to take the initiative and allocate resources toward threat modeling technologies. Simulating real-world conditions and conducting rigorous security testing before deployment can help alleviate risks. Additionally, fostering communication between technologists and security experts is essential to identify blind spots and enhance overall safety.

Implementing comprehensive cybersecurity measures in the face of rapidly developing technologies is a daunting task. The exponential growth of cutting-edge tech means that vulnerabilities multiply faster than protections can be put in place. Even when robust security measures are included in initial designs, maintaining strong defenses across an expanding threat landscape presents huge difficulties. Among other technologies, the projected increase in Internet of Things (IoT) devices—to over 30 billion by next year—will further strain security teams, even those with ample funding. The nature of emerging technologies, which prioritize rapid deployment and adoption, often clashes with the need for meticulous and extensive security controls.

To tackle these challenges effectively, security measures must be flexible and automated. They should be capable of intelligently identifying and responding to threats in a vast and ever-changing environment. Companies encounter difficulties in implementing comprehensive controls before it becomes too late. The very features that make these technologies transformative, such as hyperconnectivity, automation, data proliferation, and complexity, create digital attack surfaces that can be exploited.

For example, the ultra-low latency and high bandwidth of 5G networks enable new classes of applications, but also increase vulnerability to jamming, intercepts, and injection attacks. The interconnected nature of the Internet of Things allows for remote manipulation of physical objects in ways that legacy devices did not permit. Quantum computing has the potential to disrupt cryptography and render current public key infrastructure obsolete.

These seismic shifts present opportunities for unforeseen threats that few experts can anticipate before attacks occur. Moreover, the secretive nature of numerous newer technologies conceals their weak points from scrutiny, leaving many organizations unprepared for new forms of exploitation. As positive capabilities increase with each technological advancement, so too do the expanding possibilities for misuse.

To stay ahead of these novel threats, it is essential to research and learn how innovations can be weaponized. Equally crucial are implementing resilience through design principles and employing multivariate security controls. As long as we rely on emerging technologies, we must remain vigilant against emerging attack vectors. There is no singular solution; rather, the problem requires an enduring commitment to match ingenuity with proactive investment in protections.

As earlier mentioned, collaboration among various stakeholders, including governments, tech companies, academics, and cybersecurity experts, is vital to addressing these issues. By sharing knowledge, expertise, and resources, these groups can develop comprehensive security strategies that account for the unique capabilities and risks associated with newer cyber tools.

While the cybersecurity risks we've discussed apply to various emerging technologies, it is beneficial to delve deeper into those that impact the Internet of Things (IoT), cloud computing, 5G networks, biometric authentication, and blockchain specifically. As these technologies continue to evolve and expand globally, robust security strategies tailored to their individualities will be vital in realizing their full potential while minimizing unintended harm. By examining prominent features, inherent vulnerabilities, and recommended safeguards for each area, we can distill valuable lessons and principles to guide our cyber defense strategies.

The Internet of Things (IoT) serves as a prime example of the need for security measures tailored to newer technologies. IoT encompasses a vast network of interconnected smart devices embedded with sensors, software, and connectivity points. IoT devices, such as smart homes, wearables, and industrial controls, offer convenience through automation and data collection.

However, this extensive connectivity also introduces dangers that require careful consideration and mitigation.

Numerous IoT devices lack essential security features such as encryption, password protection, and secure boot capabilities. The limited resources available on compact IoT devices also restricts the implementation of advanced cybersecurity controls. These devices' vulnerability is further exacerbated by their global accessibility through the internet, as demonstrated by the Mirai botnet's use of unsecured IoT devices for widespread DDoS attacks.

To ensure the security of IoT equipment, it is crucial to incorporate security by design principles during their development, conduct thorough testing of vendors' cybersecurity claims, and establish clear segmentation between IoT networks and other critical systems. As IoT adoption continues to accelerate, organizations must evaluate their risk tolerance and prepare for large-scale automation.

Cloud computing is another emerging technology that necessitates careful evaluation. It offers convenient on-demand access to scalable computing resources via the internet. Public cloud services enable organizations to reduce infrastructure costs by utilizing shared resources from providers. However, this convenience also introduces new data security challenges.

When storing data in the public cloud, organizations relinquish physical control over it. Therefore, it's crucial for cloud customers to have a comprehensive understanding of the system and correctly configure strict security controls to safeguard their cloud workloads. The 2019 breach of sensitive customer data in AWS by Capital One is a reminder of how misconfigurations can expose critical assets.

In addition to misconfigurations, other cloud risks include unauthorized access through stolen credentials, data leaks during transmission or storage, and the potential for compromised insider access on the cloud provider's side. To lessen these risks, robust encryption, privileged access management, governance frameworks, and monitoring tools are essential.

5G cellular networks, a third emerging technology, promise significantly faster connectivity speeds, lower latency, and increased capacity compared to 4G networks. These advancements broaden the range of possible applications and functionalities for wireless devices. However, it's important to note that 5G also amplifies the potential for cyberattacks due to the introduction of new protocol vulnerabilities.

The 5G core's reliance on virtualization and third-party integrations makes it predominantly software-defined. This increased dependence introduces risks if adequate isolation is not enforced. The complex signaling protocols used in the 5G core have management interfaces exposed to the internet, rendering them vulnerable to attacks like jamming, intercepts, and network injection, which can bypass traditional defenses. To counter these risks, safeguards such as encryption, identity and access management, and segmentation of public and private network portions must be implemented.

Biometric authentication methods, such as fingerprint, facial, and iris recognition, offer a more secure and user-friendly approach to controlling user access based on unique biological attributes. But despite their benefits, there are concerns about biometric data breaches, presentation attacks using spoofed fingerprints, and the irrevocability of biometric credentials once compromised.

Unlike passwords and tokens, biometric identifiers cannot be easily changed if stolen or leaked. Moreover, biometric data may reveal sensitive medical conditions. Multifactor authentication and liveness detection can help address these concerns, while securely storing sensitive biometric data in hardware-backed vaults, as opposed to traditional databases, can bolster security.

Blockchain, a recent addition to the list of emerging technologies, provides decentralized and distributed ledgers for transparent and immutable data management and transactions. However, blockchain ecosystems come with inherent risks, including code vulnerabilities within blockchain platforms and smart contracts that support various functions. If flawed code results in asset loss or leakage, the damage cannot be easily undone. The management of crypto wallets and keys presents significant challenges, as evidenced by substantial thefts in recent years due to lost keys or hacks. Although the blockchain itself may be secure, it is necessary to conduct rigorous auditing and testing of endpoints and applications built upon this technology.

As emerging technologies rapidly evolve and become more dynamic, it is important to have proactive, flexible cybersecurity strategies in place to defend them. One such strategy is the shift to design-first security, which, as discussed previously, involves integrating security measures into the initial design and development stages of emerging technologies. This ensures that the foundational protections can evolve along with the technology, saving technical debt and avoiding the need for complicated retrofitting later on.

Alongside design-first security, implementing basic security hygiene practices is also essential. Robust encryption and access controls play a critical part in safeguarding sensitive data. By pervasively encrypting data, implementing role-based access controls, and utilizing multi-factor authentication, organizations can mitigate the dangers introduced by emerging technologies and establish a strong foundation of resilience.

Continuous risk assessments are another important component of effective cybersecurity strategies. Regularly evaluating threats and vulnerabilities provides valuable insights that can be used to strengthen defenses before incidents occur. By leveraging focused cyber threat intelligence, conducting penetration testing, and simulating attacks, organizations can maintain up-to-date visibility of weak points as technologies evolve.

Investment in expertise and training is crucial for securing innovation. Cultivating specialized skills, knowledge, and experience establishes the human firewall needed to protect emerging technologies. Education, certifications, exercises, and incentives can empower security teams to proactively address threats by focusing on the unique attack surfaces of emerging technology.

Adaptable security frameworks that can rapidly evolve to tackle novel threats and attack methods are likewise vital. Built-in flexibility must be added to match the dynamic nature of emerging technologies and effectively handle unforeseen risks.

Close cross-functional partnerships among security, technology, and development teams enable comprehensive perspectives on risks and unified priorities, facilitating the integration of security early in technology lifecycles. This collaborative approach ensures that security is a fundamental consideration throughout the development process, resulting in more robust, secure emerging technologies.

As cutting-edge tech continues to shape our world, it is imperative to maintain a vigilant, balanced approach to cybersecurity. By implementing proactive preparation and adaptable best practices, technology breakthroughs can be fortified against inevitable threats.

Presently, the United States is engaged in a race to bolster its defenses against nation-state actors vying for supremacy in areas such as artificial intelligence and quantum technology. These domains present special challenges with significant consequences. As the world becomes increasingly interconnected, the importance of securing new technologies cannot be overstated, particularly in the context of maintaining national security and global stability.

In the realm of artificial intelligence, nations like China have made substantial progress, investing heavily in research and development. AI has the potential to revolutionize industries from healthcare to transportation, and its applications in defense and cybersecurity are equally transformative. A nation with advanced AI capabilities could gain a strategic advantage by automating cyberattacks, enhancing surveillance, and developing autonomous weapons. To counter these threats, we must prioritize investment in AI research and foster collaboration between academia, industry, and government to develop innovative AI-based cybersecurity solutions. By doing so, we can uphold our position as a global leader in AI while ensuring the technology's safe and ethical development.

Quantum technology is another domain where we must keep pace with global advancements. Quantum computing, for example, has the potential to solve complex problems far beyond the capabilities of classical computers. This could be used to decrypt sensitive information, threatening national security and critical infrastructure. As nations like China, Russia, and the EU ramp up their quantum research efforts, we must accelerate our own programs and develop countermeasures, such as post-quantum cryptography, to protect our digital assets. Investing in quantum-resistant security measures will safeguard our data and maintain our technological edge in the face of rapidly advancing quantum capabilities.

Time is of the essence, and we as a nation must prioritize our cyber defenses to keep up with the ever-evolving threat landscape. Emerging technologies like AI and quantum computing have the potential to reshape global power dynamics, and those who master these domains will have significant advantages. As such, it is crucial to approach the development and deployment of these technologies with careful consideration and foresight. By prioritizing security and fostering collaboration between key stakeholders, the US can create an environment where emerging technologies thrive securely—ultimately enabling progress while maintaining the safety and well-being of its citizens and infrastructure.

In the coming chapters, we will delve deeper into these technologies. For now, remember that the unique capabilities that make newer technologies revolutionary also introduce unfamiliar, hard-to-predict security risks. Each technological advancement fundamentally alters the threat landscape, creating opportunities for attacks through new vectors that were previously nonexistent.

Chapter 16

THE RACE TO QUANTUM ENCRYPTION

The need for robust cybersecurity measures has never been greater. As explored in the previous chapter, we are in an era where emerging technology also presents us with sophisticated cyber threats. While technology continues to advance, we must constantly adapt and urgently seek new strategies that can keep pace with this exponentially evolving threat landscape.

We are rapidly approaching a pivotal quantum inflection point, where advanced quantum technologies could fatally crack even our best classical cryptography standards. If this happens, it will leave top secret, classified information tragically vulnerable to malicious theft and devastating attacks. Quantum computing holds the potential to upend everything we've come to rely on for data encryption and widely believed unhackable communications.

One area that could revolutionize modern cybersecurity is quantum encryption. This kind of encryption harnesses the principles of quantum mechanics to provide an unprecedented level of security for transmitting sensitive data. Unlike traditional methods, which may become vulnerable to hacking once quantum computers arrive, quantum encryption offers astonishing levels of near-unhackability. This technology has the ability to protect governments, organizations, and individuals from cyber threats and ensure the confidentiality of their information.

Many current classical encryption methods, such as AES, RSA, and SSL, face impending risks from quantum computers that could one day crack them open. These encryption algorithms are widely used and central in securing digital communications today. AES (Advanced Encryption Standard) is a symmetric encryption algorithm used to protect confidential data in various applications, including government and military communications. RSA (Rivest-Shamir-Adleman) is an asymmetric encryption algorithm commonly used for secure data transmission and digital signatures. SSL (Secure Sockets Layer) and its successor, TLS (Transport Layer Security), are cryptographic protocols that provide secure communication over computer networks, often used for web browsing and email.

One of the main components of quantum encryption is quantum key distribution (QKD). QKD leverages the unique properties of quantum bits (qubits) to create cryptographic keys. The inherent uncertainty in measuring quantum states makes it virtually impossible for an adversary to intercept the keys without being detected. As a result, quantum encryption can offer unparalleled protection against eavesdropping and other forms of cyberattacks.

QKD enables the secure exchange of encryption keys between two parties using quantum principles. Unlike classical encryption methods that rely on mathematical algorithms, QKD utilizes the properties of quantum mechanics to ensure the absolute security of the encryption keys.

In QKD protocols, the exchanged encryption keys are generated based on the quantum states of individual particles like photons or electrons. These mechanically quantum entangled particle keys are then transmitted from one party to another over a dedicated optical fiber or free space quantum communication channel encoded within laser beams. Any attempt at intercepting or eavesdropping on this quantum channel transmission by an external actor would disturb these fragile quantum states through the fundamental act of measurement at this microscopic scale.

This unavoidable disturbance would immediately alert both legitimate parties to the presence of an unauthorized third intruder trying to steal a copy of the keys over an insecure quantum communication channel. At that point, the compromised keys would simply be discarded and never used. In theory, this "observation disturbs the system" phenomenon makes directly stealing quantum bit encryption keys in transit virtually impossible. Even one stolen bit leaks information due to fragile entanglement, providing mathematically unconditional security.

Contrast this with how all classical encryption methods rely on mathematical algorithms and computational power. While these methods have held up reasonably well against current technologies, they are vulnerable to attacks from increasingly powerful computers and sophisticated hacking techniques. Given enough time and resources, these encryption methods can be broken, compromising the security of sensitive information.

A patient and well-resourced adversary or state could monitor traffic flows and secretly record scrambled communications today, planning to unlock them decades later. Only by transitioning toward fundamentally unbreakable quantum encryption can these existing gaps be closed. Quantum key distribution offers resilient protections that will last well into the uncertain quantum future ahead, rather than eroding over time. It sidesteps this great vulnerability that threatens tomorrow's data by perfectly securing today's exchanged keys through intrinsic properties of entanglement.

The unique properties of quantum encryption have the potential to secure a wide range of applications, from nuclear launch codes to bank accounts. The unbreakable nature of quantum encryption ensures that even with powerful quantum computers of the future, the encryption remains secure. This is crucial in an era where cyber threats are becoming more sophisticated and widespread.

The ability to secure everything, from critical infrastructure to personal data, has significant implications for national security, economic stability, and individual privacy. With quantum encryption, the risk of data breaches, identity theft, and unauthorized access to classified information can be greatly mitigated. It provides a level of security that is unparalleled by any classical encryption method.

However, it is important to note that while quantum encryption can protect the exchanged encryption keys, it does not provide a solution for securing the entire communication system. Other threats, such as hacking of the software or hardware infrastructure, still need to be addressed. Nonetheless, the introduction of quantum encryption brings us one step closer to achieving truly safe, unhackable communication systems.

Both the United States and China recognize the strategic advantage that quantum encryption holds in the realm of cybersecurity. As a result, both countries have invested billions of dollars in quantum research and development, as well as the necessary infrastructure to support it. The focus is on achieving quantum-resistant cryptographic capabilities to protect sensitive information from any future threats posed by quantum computers.

China, in particular, has made significant progress in the field of quantum encryption. The country has achieved several milestones, including the launch of the world's first quantum satellite, known as Quantum Experiments at Space Scale (QUESS), in 2016. This satellite enabled the testing of quantum encryption over long distances and showcased China's commitment to quantum technology.

China has also poured substantial resources into developing top talent in the field of quantum research. They have established dedicated institutions, such as the Hefei National Laboratory for Physical Sciences at the Microscale, to foster advancements in quantum science and technology. Concurrently, Chinese tech giants like Alibaba and Huawei have taken an active role in advancing practical applications of quantum encryption, exploring its commercial potential, and contributing to the overall growth of the nation's quantum capabilities.

The rapid progress and achievements China has made in quantum encryption have raised concerns about the implications for US national security. There is worry that China's quantum lead could undermine the United States' position in cybersecurity and intelligence gathering. China's advancements in quantum encryption technology might provide them with an advantage—the ability to protect sensitive information while compromising the security of other nations.

However, the development and implementation of quantum encryption will certainly not be limited to any single country. An intensifying and dangerous competition already rages between the United States and China, making them a focal point in the emerging quantum arms race. Both countries have been investing heavily in research and development to gain the upper hand in this field. Now, with geopolitical pressures mounting, the race for undisputed quantum supremacy has exponentially raised the stakes.

While the United States has historically been at the forefront of technological innovation, China's focused investment in quantum research has led to remarkable advancements. Its substantial strides in quantum communication networks, quantum satellites, and quantum computing have narrowed the gap between the two nations in this strategically important domain.

Given the importance of cybersecurity in an increasingly interconnected world, China's growing capabilities in quantum encryption have prompted questions about the vulnerabilities and risks this could create for other countries. As the US seeks to maintain leadership in technology and ensure the security of our digital infrastructure, understanding the potential impacts of China's quantum advancements becomes crucial.

One key concern is the that quantum encryption may disrupt existing security protocols and provide an upper hand to those who master the technology first. If China achieves a significant lead in quantum encryption, it could compromise the security of our communications and sensitive data. Superiority in this field could alter power dynamics in both economic and military spheres.

Another worrying aspect is the vulnerabilities that could stem from China's dominance in quantum encryption technology. Should another country or organization gain access to China's quantum encryption techniques, they might compromise the security of various entities and obtain unauthorized access to confidential information. Such a situation would pose a considerable threat to national security and the stability of global networks.

As the US and China engage in this escalating quantum arms race, the implications extend far beyond the realm of cybersecurity. Strategic military advantages, economic competitiveness, and technological supremacy hang in the balance. The competition to achieve quantum supremacy is becoming increasingly fierce, and the outcomes of this race will have far-reaching implications, not only for cybersecurity but also for various other sectors, including communication, information technology, and scientific advancements.

China’s growing capabilities in quantum encryption could tip the balance of technological power, leading to a significant shift in the global landscape. This development prompts questions about the US's ability to maintain our competitive edge, protect our economic interests, and safeguard national security in an increasingly complex digital world.

In light of these concerns, it becomes evident that we must urgently understand and address the implications of China's quantum advancements. Fostering a spirit of collaboration and responsible innovation is essential as we work towards ensuring the security and stability of our interconnected world. By promoting open standards for quantum encryption and advocating for responsible practices, we can help maintain the integrity of the digital landscape while working with international partners to ensure that advances in quantum technology benefit the global community without exacerbating existing geopolitical tensions.

However, China's rapid progress across various technological domains, such as mobile payments, 5G telecommunications, virtual reality, genetic engineering, and now quantum encryption, signals a potential shift in the long-held competitive advantage that has powered the economic and military dominance of the United States. This progress could pose the first real threat to this dominance since the Soviet Sputnik launch, with far-reaching consequences that extend beyond trade deficits to encompass global influence and even notions of national identity.

As we consider the implications of China's advancements in quantum encryption, it is crucial to recognize the impact it may have on the broader technological landscape. We must continue to invest in research and development, as well as cultivate partnerships with international allies, to maintain our leadership in technology and security. By doing so, we can help shape the future of quantum encryption in a way that protects the interests of Americans and the world, creating a culture of innovation and cooperation in the face of evolving global challenges.

It is essential for stakeholders and other nations to closely monitor China's progress in quantum encryption and take necessary measures to address any potential gaps in their cybersecurity strategies. As quantum encryption continues to advance, we must establish robust international partnerships and frameworks to ensure the security and stability of global networks and information exchange.

The cybersecurity imbalance that heavily favors China poses significant challenges for other nations and has resulted in economic losses, particularly in the fields of communications and finance. This imbalance also raises enhanced threats to critical infrastructure and national interests, as China's advancements in quantum encryption technology give it an advantage in targeting and exploiting vulnerabilities.

One concerning aspect of the imbalance is China's ability to shape global tech standards and regulations. With its growing influence in the global tech landscape, China can potentially exert significant control over the rules and protocols that govern the use of technology worldwide. This can lead to standards that may not align with the interests of the United States and other nations, creating a possible risk to global cybersecurity.

Furthermore, China's advancements in this field give it a greater ability to break the security systems of networks and organizations. This increases the risk of cyberattacks, data breaches, and other forms of malicious activity targeting US interests.

One key implication is the emerging threat of quantum-enhanced ransomware. With China's advancements in quantum computing, cybercriminals may leverage this technology to develop more sophisticated ransomware attacks. Quantum computational power could break encryption systems more easily, posing a greater risk to individuals, organizations, and even critical infrastructure.

This leads to another implication: the need for quantum-resistant security measures. As China continues to develop its quantum computing capabilities, there is a growing urgency to develop encryption methods that are resistant to quantum hacking. Collaborative efforts between the US, China, and other nations are essential for researching and implementing quantum-resistant security measures to ensure the long-term protection of sensitive data and critical systems.

The significance of building trust and fostering open communication channels between the United States and China cannot be overstated in this pursuit of global cybersecurity. Despite our historical rivalries and differences, a strong foundation of trust is essential for meaningful cooperation in addressing the cybersecurity imbalance and advancing quantum encryption technology. By encouraging open dialogue, transparency, and mutual understanding, both nations can create an environment conducive to sharing knowledge, resources, and expertise.

To facilitate this, the establishment of joint research initiatives, academic exchanges, and industry partnerships between the United States and China can be highly beneficial. These collaborative relationships can create opportunities for experts and researchers from both nations to work together, share insights, and develop innovative solutions to address the cybersecurity challenges posed by quantum encryption.

That said, the need to balance cooperation with safeguarding national security interests is paramount. While teamwork is crucial, it is also essential that nations remain vigilant in protecting their critical infrastructure, intellectual property, and sensitive information. Establishing clear guidelines, setting boundaries, and implementing robust protections can ensure that national security interests are not compromised while we pursue the shared goals of advancing quantum encryption technology and improving global cyber safety.

In conclusion, the cybersecurity imbalance resulting from China's advancements in quantum encryption presents both trials and opportunities for the United States and the global community. By recognizing the need for collaboration, trust, and investment in joint efforts, nations can collectively address the risks associated with quantum encryption. In doing so, they can pave the way for a more secure and interconnected digital world, where quantum encryption serves as a force for stability, prosperity, and global cooperation.

Chapter 17
THE AI REVOLUTION

We are closer now than ever to a future where intelligent machines interact with humans as routinely as other people do—a time when artificial intelligence systems will collaborate with human cybersecurity analysts to swiftly identify and neutralize threats. This future is not a distant dream, considering the accelerating pace at which disruptive technologies are emerging. However, it is an uncertain prospect, given that it's partly driven by nations racing to secure a prominent position in the evolving landscape of technology.

Artificial intelligence is one of the emerging technologies gaining preeminence today. It is focused on the simulation of human intelligence into machines, a feature which would allow machines to think and learn like humans. The limitations in present-day computing systems necessitate the further development of AI that can perform tasks that would normally require human intelligence. These tasks may include understanding natural language, recognizing patterns, solving problems or making decisions.

In modern technology, AI has become a crucial component that is transforming various domains. One of its major applications is in the creation of chatbots. These are computer programs designed to simulate human conversation. AI-powered chatbots can understand natural language and respond to users' queries in a conversational style. They are used in customer service for providing instant assistance and support, and for handling large volumes of inquiries. Their effectiveness helps to free up human agents for more complex tasks.

Another significant application of AI is in self-driving cars. AI algorithms, combined with sensors and cameras, enable these vehicles to perceive and interpret their surroundings, make decisions in real time, and navigate autonomously. These self-driving cars have the potential to improve road safety, reduce accidents caused by human error, and provide accessible transportation options for individuals who are unable to drive themselves.

Facial recognition technology also uses AI algorithms, in this case to analyze and identify unique facial features, allowing for the identification and verification of individuals. Facial recognition systems are used in security and surveillance, access control for devices and buildings, and even in social media platforms for photo tagging and augmented reality experiences.

As AI gains more interaction with other technologies and industries, it has begun to play a significant role in healthcare. Medical professionals now use AI algorithms to analyze complex medical data, such as medical images, patient records and research papers. The use of AI enables more accurate and efficient diagnostics, personalized treatment recommendations, and the identification of risks or abnormalities. These AI-based systems can assist in drug development, patient monitoring, and the early detection of diseases, ultimately improving patient outcomes and reducing healthcare costs.

Other domains benefiting from AI include finance, where AI algorithms are used for fraud detection, algorithmic trading, and risk assessment; manufacturing, where it's used for quality control, predictive maintenance, and automation of manufacturing processes; and entertainment, where it's employed for content generation, content recommendation, and personalized marketing. Education also benefits from AI through intelligent tutoring systems, adaptive learning platforms and automated grading. In short, it has become deeply entwined with almost all industries.

As an emerging technology, AI holds the capacity to revolutionize a multitude of sectors, including cybersecurity. In this particular context, it can serve as both a powerful tool and a possible threat. On one hand, it can enhance cybersecurity by enabling advanced threat detection and response mechanisms. AI algorithms can analyze vast amounts of data, identify patterns, and detect anomalies to flag potential security incidents. This capability empowers organizations to more effectively neutralize cyber threats, thereby improving their overall security posture.

On the other hand, AI can also pose challenges for cybersecurity. Hackers and cybercriminals can leverage it to launch more sophisticated and automated attacks. AI-powered malware and phishing can more easily deceive users, making them harder to detect and mitigate. AI can also be used to amplify the scale and speed of cyberattacks, thereby overwhelming security systems and infrastructure.

Because a nation's strength is partly a function of its cybersecurity posture, AI presents both opportunities and challenges to not only cybersecurity, but national security as well. This technology can enhance the capabilities of security agencies and military forces by providing more advanced surveillance systems, intelligent threat detection and autonomous defense systems. It can enable faster and more accurate decision-making in critical situations, potentially saving lives and resources.

But AI's integration into national security also introduces unique pitfalls. As these systems become more prevalent in sensitive domains, the prospective of cyberattacks targeting them becomes a pressing concern. Malicious actors may seek to exploit vulnerabilities in AI technologies to manipulate or sabotage them, which could have serious consequences for national security. Furthermore, the incorporation of AI into autonomous weapon systems raises complex ethical and legal issues concerning control and accountability in their deployment and usage.

While this emerging technology amplifies many opportunities and risks, it holds tremendous potential for greater advancement into other areas of our daily lives. This explosion of AI advancements begs the question: Where did it all begin? Let's rewind and trace AI's remarkable evolution, connecting historical milestones to their impact on the world today.

The origin of AI can be traced back to ancient times, with philosophers and mathematicians pondering the possibility of creating artificial beings capable of human-like thought. However, it wasn't until the 20th century that significant progress was made.

In the 1940s, the invention of the first programmable computer, the Electronic Numerical Integrator and

Computer (ENIAC), was a pivotal catalyst for AI. The ENIAC provided researchers with a platform to explore the possibility of automating tasks that require human intelligence.

Throughout the 1950s and 1960s, researchers from various disciplines, including mathematics, logic, psychology, neuroscience and computer science, contributed to AI's progress. Influential figures like Alan Turing and John McCarthy laid the foundations by posing fundamental questions and proposing groundbreaking ideas.

The arrival of artificial neural networks in the 1980s proved a significant milestone in AI development. These networks, inspired by the complex connections in the human brain, allowed for the processing of complex information and ultimately led to breakthroughs in areas like speech and image recognition.

As the field improves, researchers are now facing the challenge of adversarial AI. The adversarial AI's attacks and defenses aim to manipulate AI systems by exploiting vulnerabilities. These include techniques like generating deceptive inputs or subtly altering data to fool AI algorithms. Mitigating these challenges requires robust security measures and constantly evolving AI algorithms.

From ancient philosophical musings to modern breakthroughs, the diverse influences and disciplines of mathematics, logic, psychology, neuroscience, and computer science have all shaped AI's trajectory. Today, it continues to push boundaries, changing our world and driving innovations that were once the stuff of science fiction.

Despite the long path it took, the goals for developing AI have remained the same, revolving around problem-solving, learning and adaptation. AI has certain components that work together to make all this possible. Each part contributes to the goals in different ways, enabling AI systems to tackle complex tasks and improve their performance over time. Let's explore some key features of this technology and how they contribute.

One crucial feature of AI is machine learning. Think of it as learning to cook through trial and error. Imagine you're trying to make the perfect chocolate chip cookie. At first, you follow a recipe, bake the cookies, and taste them. Maybe they turn out too sweet or not chewy enough. You make adjustments, maybe reducing the sugar or modifying the baking time, and bake another batch. Through this iterative process, you learn from your previous attempts and improve your cookies over time.

Similarly, machine learning enables AI systems to learn from data and improve their performance through iterations. Instead of working with explicitly programmed rules, they learn patterns and relationships from large amounts of data. For example, in spam email detection, a machine learning algorithm analyzes countless emails and learns to distinguish between legitimate and spam emails based on patterns it discovers in the data.

Natural language processing is another essential component of AI. Imagine having a conversation with an AI that is similar to chatting with a friend. Natural language processing enables AI systems to comprehend and generate native-sounding human language. For example, chatbots or virtual assistants like Siri or Alexa utilize natural language processing to understand spoken or written queries and provide relevant responses.

Computer vision allows AI systems to "see" and interpret visual data as humans do. Think of it as the ability to recognize objects in images or videos. For instance, facial recognition technology uses computer vision algorithms to detect and identify faces in photographs, video footage or other visual feeds.

Neural networks also play a significant role in AI. As earlier stated, they were inspired by the complex connections found in the human brain. These networks are a type of machine learning algorithm that learns hierarchical representations of data. Just like a brain has interconnected neurons, a neural network has interconnected nodes called artificial neurons. These nodes are involved in organizing and processing information. Neural networks have proven incredibly effective in tasks like image classification, speech recognition and natural language processing.

Reinforcement learning is another element, which involves training AI systems through a reward-based system. Imagine teaching a dog new tricks. When it performs the desired action, you provide positive reinforcement, like treats or praise. Over time, the dog learns to associate that action with the reward and performs it consistently. Reinforcement learning follows a similar principle, where AI agents learn optimal actions in a given environment by receiving rewards or penalties based on their performance.

Expert systems are an aspect of AI designed to mimic the expertise and decision-making processes of human experts.

They encode human knowledge and rules into a computer system, enabling it to make informed decisions in specialized domains. These systems have been applied in medicine to help doctors make accurate diagnoses based on patient symptoms.

Each of the components we've discussed contributes to the broad goals of problem-solving, learning, and adaptation. Together, they allow AI systems to tackle complex tasks, understand and generate natural language, interpret visual data, learn from experiences, and make informed decisions based on specialized knowledge.

When examining AI components, it helps to classify AI systems based on criteria like functionality, capability and learning mode. Understanding these classifications provides deeper insight into the diverse AI landscape. Now, let's explore three distinct AI categories: weak AI, strong AI, and artificial superintelligence.

Weak AI, also known as narrow AI, is designed to excel in specific tasks or domains. It can outperform humans in those areas but lacks general intelligence. For instance, a chess-playing AI may consistently defeat human opponents at chess, but it cannot perform complex tasks outside the game. This category of AI is prevalent in various applications, including voice assistants like Apple's Siri or recommendation systems like Netflix's algorithm.

On the other hand, strong AI, also known as general AI, possesses broader intelligence and can perform tasks at the human level or beyond across multiple domains. It exhibits cognitive abilities resembling those of a human in a variety of areas. Strong AI would be comparable to an AI system that reads, comprehends, and converses in multiple languages. It may even show creativity and adapt to new situations. Achieving strong AI is a significant challenge that researchers are actively pursuing.

Artificial superintelligence takes the concept further, referring to AI systems that surpass human intelligence by a wide margin. This category of AI embodies hypothetical abilities that could outmatch human comprehension and problem-solving capabilities in all domains. While artificial superintelligence remains speculative, it sparks both excitement and concern due to its potential impact on society, ethics and existential risks.

Looking at the current state of AI research and development, recent breakthroughs are pushing boundaries further. The Turing test, proposed by Alan Turing, assesses whether a machine's responses are indistinguishable from a human's during a text-based conversation. The AlphaGo challenge showcased deep reinforcement learning's power, with Google's AI defeating world champion players in the complex game of Go. The GPT-3 model by OpenAI demonstrated impressive natural language processing capabilities by generating human-like text. Notably, OpenAI's Codex showcases the future of AI in programming, enabling automatic code generation and assisting developers in their coding tasks.

So how do recent breakthroughs impact cybersecurity and national security? AI promises to improve these areas significantly. For instance, it improves threat detection by analyzing vast amounts of data, identifying patterns and flagging possible security breaches. Systems like IBM

Security QRadar, Guardium, and Security Verify leverage AI to strengthen cybersecurity defenses and shield sensitive information.

However, AI also introduces potential threats. Adversaries can exploit it for cyberattacks, utilizing AI-driven algorithms to generate sophisticated malware or bypass existing security measures. Imagine malicious AI that autonomously searches for vulnerabilities and launches targeted attacks. Ethical and legal considerations are paramount in addressing these challenges, including concerns about transparency, accountability, privacy, bias in AI algorithms, and regulations on AI in cybersecurity.

To protect the future of cybersecurity and national security, collaboration and coordination among stakeholders are essential. Governments, businesses, academia, and civil society must work together to establish ethical frameworks and regulations that balance innovation with safeguards against risks. Striking this delicate balance will foster trust in AI-driven cybersecurity and ensure a secure and prosperous future.

As we conclude this chapter, we acknowledge the significant role AI plays in shaping our digital landscape. In the following chapters, we will further explore the world of "GenAI in Cybersecurity," examining AI's evolving challenges and possibilities, where AI emerges both as a cybersecurity threat and solution.

Chapter 18

THE RISE OF GENAI IN CYBERSECURITY

In the previous chapter, we delved into the world of artificial intelligence and its profound impact on cybersecurity. We discovered that AI, while offering opportunities for enhanced protection, also presents new challenges as adversaries harness its power for sophisticated attacks. Now, we turn our attention to the emergence of generative AI (GenAI) and its implications for cyber- and national security.

Imagine GenAI as a skilled engineer, crafting innovative solutions by adapting existing technologies into advanced defense systems. GenAI leverages algorithms and diverse data sources, such as text, images, audio, and video, to generate novel outputs. This generative capability has the potential to revolutionize various domains, from intelligence analysis and mission planning to training simulations, opening unprecedented opportunities for data analysis and decision-making.

However, the rise of GenAI also introduces new risks. Adversaries could use this technology to create malicious content, fueling more advanced social engineering attacks or overwhelming cyber defenses with waves of fake threats. As we explore the power of GenAI, we must also address critical concerns regarding policy, legality, and ethics, spanning topics like synthetic media, data privacy, and countermeasures against AI-driven cyber threats.

At GenAI's core lies the foundation of machine learning, a branch of artificial intelligence enabling computers to learn patterns and make decisions without explicit programming. Deep learning, a subset of machine learning, plays a pivotal role. Like a soldier honing their ability to identify dangers, deep learning models process data in layers, each extracting and interpreting different features to recognize possible risks. This pattern recognition empowers deep-learning models to generate new data, aiding in predicting and combating emerging cybersecurity threats.

GenAI employs various techniques to unlock its creative potential, including generative models explicitly engineered to produce new data. These models can create a range of synthetic outputs, from images and videos to text and even code. Among these models are generative adversarial networks (GANs), variational autoencoders (VAEs), and generative pre-trained transformers (GPTs).

GANs, for instance, operate like two military strategists collaborating to develop realistic training simulations. They comprise two neural networks—the generator and the discriminator. The generator's role is to create synthetic data resembling real-world scenarios, accurately simulating likely threats. In contrast, the discriminator tries to distinguish between real and generated scenarios. As the generator improves at deceiving the discriminator, it becomes increasingly adept at producing realistic and valuable training simulations.

In this chapter, we'll delve deeper into the mechanics of GANs, VAEs, and GPTs, exploring how they contribute to the rapidly evolving landscape of GenAI. As we do so, we'll uncover insights into their potential applications and the challenges they present, both in the realm of cybersecurity and beyond.

As we march forward, let's first enter the world of transformers, the linguistic commandos of the neural network realm. Picture transformers as skilled storytellers, adept at weaving intricate narratives with their words. These neural network architectures excel at processing sequential data, such as words in a sentence or paragraphs in a document. They employ attention mechanisms to prioritize the importance of individual words, effectively capturing the nuanced relationships within the data. This unique ability enables transformers to generate coherent and natural language text, much like a seasoned orator.

Now, let's shift our focus to variational autoencoders, or VAEs, the artistic commandos of the AI world. VAEs are like skilled painters with a special twist. They possess the exceptional ability to capture the essence of a given dataset, be it images or other visual data, by deciphering the underlying distribution of information. This talent allows them to produce novel, synthetic examples that bear a striking resemblance to the original inputs while introducing subtle variations, offering a spectrum of creative possibilities that push the boundaries of imagination.

With these innovative techniques in mind, let's explore some of the most popular GenAI platforms and tools that harness the power of these concepts. Take, for instance,

ChatGPT, an AI language model that serves as a linguistic warrior, engaging in informative and captivating conversations. This model possesses the capability to generate human-like responses and deliver relevant information based on the context of the discussion.

DALL-E2, on the other hand, takes GenAI capabilities down a visual route. This platform can create images from textual descriptions, much like a talented illustrator. By inputting a vivid description, such as "a cat with the body of a pineapple," DALL-E2 can bring that imaginative vision to life through a captivating picture.

Stable Diffusion, another powerful tool in the GenAI arsenal, leverages the power of GANs and other techniques to transform text into photorealistic images. This platform employs diffusion models to produce diverse, high-quality samples from a given starting point, opening the door to controlled exploration of the underlying data distribution and paving the way for the creation of unique and imaginative outputs.

Among other notable examples, we find GitHub's Copilot for code autocompletion and Google Gemini for conversational AI. These platforms demonstrate not only GenAI's immense promise, but also the practical applications that stem from techniques like GANs, transformers, and VAEs. They stand as a testament to the raw power and ingenuity of AI-generated content.

Turning our attention to GenAI's implications for cybersecurity, we must recognize the game-changing capabilities it brings to the battlefield of cyber defense and offense. Let's explore key use cases and potential risks of GenAI through examples across these domains.

On the frontlines of cyber defense, GenAI emerges as a strategic ally, fortifying our digital boundaries. One crucial tactic it employs is the generation of synthetic data, akin to a military drill that simulates various combat scenarios. GenAI can artificially create new malware samples, bolstering threat intelligence datasets and enhancing the detection of zero-day attacks. This generative power equips cybersecurity systems with invaluable synthetic data, allowing them to better identify deviations from normal patterns and swiftly detect possible dangers lurking in the digital shadows.

In the face of cyber warfare, GenAI proves its value by analyzing vast amounts of data to quickly spot threats, much like an intelligence analyst sifting through mountains of information. It also assists in encryption, safeguarding sensitive information from unauthorized access. It facilitates the deployment of cutting-edge data privacy techniques, such as differential privacy and federated learning, which offer secure data insights without compromising the integrity of the raw data.

Lastly, GenAI serves as an innovation incubator, nurturing the development of novel cybersecurity solutions. It empowers researchers and businesses to explore uncharted territories and devise groundbreaking strategies in the ongoing battle against cyber threats. GenAI stands as a critical weapon in the arms race to safeguard America's digital infrastructure and protect its citizens from the looming threats of the modern age.

As we delve deeper into the potential of GenAI, it's important to acknowledge that its dual-natured capabilities can be harnessed for both defensive and offensive purposes. The same generative power that makes GenAI so valuable can be exploited for malicious ends, enabling cyber offense. For instance, malicious actors can use GenAI to produce fake content, such as deepfakes. Deepfakes are manipulated videos or images that convincingly depict a person saying or doing something they actually did not. This can have severe consequences, including reputational damage and the spreading of misinformation. It can also be used to fuel advanced social engineering campaigns.

On the offensive front, GenAI can be employed in phishing attacks, where adversaries craft synthetic emails or messages to deceive individuals into disclosing sensitive information. It's similar to an enemy using deceptive tactics to infiltrate our ranks and gain access to classified intelligence. Furthermore, GenAI can facilitate impersonation, enabling adversaries to mimic voices or create synthetic identities to mislead unsuspecting targets—similar to donning a convincing disguise to blend in and manipulate our actions.

Given these offensive use cases, the imperative for ethical frameworks and governance surrounding GenAI becomes increasingly clear. Just as a military unit operates under a strict code of conduct to maintain discipline and ensure the welfare of its members, so too must we establish guidelines for the responsible development and deployment of GenAI. These principles serve as our guiding stars, illuminating the path toward beneficial AI innovation while minimizing potential risks.

One of the cornerstones of these ethical frameworks is transparency and explainability. In the context of GenAI, this principle involves openly disclosing the origins of the data used to train the AI and providing clarity about its capabilities. This practice fosters trust and accountability, much like a commander briefing their troops on the details of a mission to instill confidence and unity of purpose. By shining a light on the inner workings of GenAI, we can proactively identify vulnerabilities and take corrective action when necessary.

Another pillar of ethical GenAI is accountability. As these systems are increasingly designed make autonomous decisions within the realm of cybersecurity, it's crucial to establish clear lines of responsibility. When the heat of battle intensifies and things go awry, we must know who is accountable for the actions of the AI. This ensures that developers, operators, and users alike are held responsible for their roles in shaping and deploying these powerful tools.

Privacy stands as a non-negotiable principle which should command our utmost respect. GenAI systems often require access to sensitive data to effectively detect threats and discern patterns. Much like a soldier swearing an oath to protect their nation, we must safeguard the personal data entrusted to us, respecting users' preferences and upholding their rights. By adhering to robust privacy protections and complying with applicable laws and regulations, we can maintain the trust of those whose data we handle.

The principle of inclusiveness calls us to address potential biases that may lurk within GenAI algorithms. Just as a diverse and cohesive unit is a force to be reckoned with on the battlefield, so too must our GenAI systems draw from diverse and representative datasets to minimize discriminatory outcomes. Through vigilant audits and evaluations, we can identify and mitigate bias, ensuring that our AI-driven decisions are fair and equitable for all.

Finally, there is the imperative for coordination among various stakeholders. Governments, businesses, academia, and civil society organizations must band together to forge ethical guidelines, regulatory frameworks, and best practices. This collaborative effort mirrors the unified front presented by allied forces, standing shoulder to shoulder to protect our shared values and way of life.

As we navigate the complexities of GenAI, these principles serve as our compass, guiding us toward responsible innovation that safeguards the interests of individuals, organizations, and nations alike. Through unity, transparency, accountability, privacy, inclusiveness, and coordination, we can harness GenAI's power to fortify our defenses and deter adversaries in the ever-evolving cybersecurity landscape.

In these pages, we have embarked on a journey through the captivating realm of generative AI and its profound influence on cybersecurity. GenAI, a trailblazing force of machine creativity, possesses the remarkable ability to produce novel data across various domains, from text and images to audio and beyond. We delved into the intricate mechanisms powering this innovation, exploring key GenAI techniques such as deep learning, neural networks, generative adversarial networks, and transformers. These groundbreaking methodologies empower GenAI platforms to synthesize data for widespread applications, including content creation, analysis, and design, transforming industries and reshaping our world.

However, the emergence of GenAI also brings forth a duality of consequences within the cybersecurity sphere. On one hand, it bolsters our defenses by harnessing synthetic data, fortifying analytics capabilities, and stimulating technological growth. Conversely, its generative prowess introduces new dangers, enabling sophisticated social engineering attacks, overwhelming defenders, and casting a shadow of doubt over the reliability of information. Our discourse highlighted the critical importance of establishing ethical guardrails and fostering collaboration among stakeholders to amplify GenAI's benefits while lessening the risks accompanying this technological leap.

As we look to the horizon, the maturing landscape of GenAI is shaped by emerging trends like explainable AI and the symbiotic partnership between humans and computers. Yet, GenAI serves as merely a stepping stone in the broader pursuit of artificial general intelligence—AI that transcends human cognitive capabilities across all domains. The global race to attain this zenith is heating up, with cybersecurity and technology leaders locked in a relentless pursuit of AI supremacy. The nation that first unlocks the full potential of artificial general intelligence stands to gain an unparalleled strategic advantage, influencing the trajectory of intelligence on Earth and beyond.

The significance of this race cannot be overstated. The choices we make today, particularly those surrounding ethics, governance, and accessibility, can steer the course of this rapidly evolving field towards prosperity or catastrophe for us as a nation. As we forge ahead, we must approach this pursuit with a heightened sense of responsibility, as well as a commitment to ensuring that the fruits of AI serve the betterment of our world and its inhabitants.

Chapter 19

THE RACE TO GENAI SUPREMACY

In the previous chapter, we explored the emergence of generative AI (GenAI) and its potential impact on cybersecurity. Though GenAI represents a significant leap in machine creativity, it currently functions as a form of narrow AI focused on specific tasks. However, the ultimate goal that technology leaders and nations are fervently pursuing is the development of artificial general intelligence (AGI)—AI capable of learning and reasoning across different domains, potentially surpassing human cognition. The race to GenAI supremacy is not just a competition to achieve AGI, but also a critical component of America's cybersecurity endgame, with the potential to shape the future trajectory of humanity.

As the urgency to secure America's digital landscape intensifies, the development of AGI becomes even more crucial. AGI's ability to revolutionize cybersecurity cannot be understated; it could detect, analyze, and mitigate threats with unprecedented speed and accuracy, transforming the way we protect our digital assets and infrastructure.

Although the concept of AGI was introduced in the 1950s, recent advancements in GenAI have brought it closer to reality. While today's most advanced AI systems can match certain aspects of human intelligence, numerous barriers involving technology, neuroscience, and computing power still stand in the way of building a general-purpose AI system that cohesively integrates these capabilities.

To overcome these obstacles and bridge the gap between GenAI and AGI, various approaches are being explored. Some are focused on scaling up machine learning models to process more data and build foundational skills, while others are simulating neuroscience processes through new hardware or combining neural networks with classical algorithms in hybrid systems. The race to GenAI supremacy is wide open, with a range of companies and startups placing their bets.

Underpinning the race to GenAI is the mounting growth of computing power dictated by Moore's Law. Each year, computing performance increases exponentially, paving the way for rapid advancements in AI capabilities. NVIDIA estimates that computing performance has grown by a factor of 4186.41% since 2012, far surpassing the progress made in the previous six decades of AI research. Additionally, technologies like quantum computing hold the potential to provide further exponential growth, propelling breakthroughs in GenAI.

Experts estimate that achieving AGI could take us anywhere from 20 to 100 years, considering the multidisciplinary knowledge required across fields like neuroscience, computer vision, and language understanding. However, access to data will play a pivotal role in this race. The nation or company that can aggregate, organize, and apply the largest datasets will have a significant advantage in developing AGI, further amplified by the use of GenAI for synthetic data generation.

The imminence of AGI raises critical questions about its potential impact on humanity. How will machine intelligence surpassing that of humans affect our future? Should access to such powerful technology be restricted? Could use of AGI lead to unintended consequences like surveillance states or totalitarian control? Balancing security, ethics, and governance will determine whether AGI leads to a technology-driven utopia or an apocalyptic dystopia.

For America, the choices made regarding GenAI today will result in one of two contrasting futures. In the responsible scenario, the benefits of AGI will be harnessed while maintaining modern human rights and democratic principles, allowing regulated AI access to leading companies and research institutions. This future entails expanding the boundaries of science under careful oversight, reminiscent of how nuclear energy was allowed to mature under civilian control.

On the other hand, a scenario driven solely by the thirst for supremacy and economic gains could lead to unfettered development behind closed doors. Power would become concentrated in the hands of a few dominant conglomerates and elite government agencies associated with defense organizations. The lack of coordination and oversight in this dystopian future would result in engineered influence campaigns, predictive policing, and AI-driven cyber warfare.

The race to GenAI supremacy is not only a national endeavor, but also a global power dynamic. The ongoing competition between the United States and China is particularly noteworthy in regards to AI, as both nations recognize the overwhelming possibilities of GenAI and AGI and are deeply involved in research and development. The outcome of this competition will have far-reaching implications for technological leadership, economic influence and geopolitical power.

As the race to achieve AGI unfolds, GenAI remains a critical step toward realizing this revolutionary milestone. The groundbreaking nature of GenAI and the pursuit of AGI have a huge influence on global power dynamics, and the decisions made today will guide the future trajectory of both AI and humanity as a whole.

Currently, the United States holds a prominent position in GenAI development. While China has made considerable progress in AI, particularly in terms of government support, infrastructure, and applied AI, the United States' substantial funding, thriving tech ecosystem, and robust academic research have allowed it to maintain a competitive edge in the GenAI niche.

Several factors contribute to America's advantage, including its leadership in cutting-edge research, access to top talent, and the presence of major US tech companies at the forefront of AI innovation. The United States government acknowledges the strategic importance of GenAI and has allocated sizeable resources to advance research and development in this field, proactively exploring its application across various sectors.

In the defense sector, for instance, GenAI offers the potential to enhance autonomous systems and decision-making processes, bolstering national security. The US leverages GenAI platforms like OpenAI Codex and DARPA's GALE to generate code, natural language, and data analysis for cybersecurity and intelligence purposes.

In the healthcare sector, GenAI can revolutionize diagnostics, treatment, and patient outcomes. The US employs GenAI tools like IBM Watson and Google DeepMind to generate medical insights based on vast amounts of data. Similarly, the finance industry benefits from GenAI's ability to develop sophisticated algorithms for risk assessment, trading, and fraud detection. GenAI systems like AlphaSense and Kensho produce financial reports, market predictions, and investment strategies based on natural language processing and data mining.

As the global competition in AI development continues to intensify, the Chinese government has made it a strategic priority for national development. As part of their frantic effort to close the gap, the state is making significant investments in research and development, while also providing substantial funding to nurture talent and accelerate technological advancements. These commitments have enabled China to make great strides in GenAI, positioning itself as a major player in the world's AI ecosystem.

One key aspect of China's approach to GenAI is its emphasis on strategic partnerships. Chinese tech companies, such as Baidu, Alibaba, and Tencent, have formed collaborations with international institutions and organizations to enhance their GenAI capabilities. These partnerships facilitate knowledge exchange, access to diverse datasets, and teamwork on cutting-edge research, thereby propelling China's work in AI innovation.

Another main element of China's approach is its dedication to AI education and research. Chinese universities and research institutes have been actively fostering talent and conducting groundbreaking research in AI, including GenAI. The government has implemented policies to attract and retain top AI talent, including encouraging students to pursue AI-related disciplines. This push for education and research has created a talent pipeline that fuels China's GenAI advancements and strengthens its position in the world of AI at large.

Chinese tech companies, particularly those mentioned earlier, have gained global influence in areas like facial recognition technology. These companies are at the forefront of research and development in facial recognition. They lead in both implementation and innovation, using GenAI platforms like Face++ and SenseTime to generate realistic and accurate facial images and features for various applications.

Their technologies have been adopted across various sectors, including law enforcement, public security, and commercial applications. The widespread use of facial recognition technology in China showcases the substantial progress the country has made in the GenAI domain. In public infrastructure, China supports AI in smart cities, using GenAI tools like Baidu Brain and Alibaba City Brain to generate data and insights for urban planning, traffic management, and environmental protection.

China's rapid progress in GenAI is evidence of its ambition and commitment to technological advancement. The combination of state investment, strategic partnerships, and a focus on AI education and research has enabled the nation to position itself as a key world player in GenAI. As China continues to make strides in this field, it further fuels the competition between them and the United States for GenAI supremacy.

While China's growth in this area is remarkable, estimating the timeline for them to achieve GenAI parity with the United States is difficult. This prediction involves an analysis of various factors, including governmental policies, investment trends, and international relations. However, geopolitical considerations and the United States' strong position in GenAI development also play a significant role. Hence, the competition for GenAI leadership between the US and China is a complex, multifaceted one. There isn't a single, clear-cut winner.

The risks associated with China achieving GenAI parity with the United States are notable. If it occurs, global economic and military power dynamics may shift. This raises concerns about ethical standards, data privacy, human rights and surveillance. The widespread adoption of GenAI, combined with China's influence, could have broad societal and governance implications.

As other countries make significant strides in GenAI development, it poses both challenges and opportunities for the US. One of the key challenges lies in striking the right balance between research and development and ethics. While it is vital to foster ingenuity and push the boundaries of GenAI capabilities, it is equally important to address ethical concerns around privacy, bias, and accountability. By advocating for robust AI governance frameworks, the US can ensure GenAI's responsible growth and deployment, thus protecting individual rights and societal values.

To maintain its position at the forefront of GenAI, the US needs to attract top talent and feed innovation. Public-private partnerships can facilitate collaboration, bringing together the expertise of academia, industry, and government. Through these combined efforts, the US can leverage varied perspectives and shared resources to push GenAI advancements further. Additionally, strategic immigration strategies can help attract talent from other parts of the world, creating a vibrant and diverse GenAI ecosystem within the country.

GenAI advancements may lead to concerns about job displacement and social inequality, and the US must proactively address these issues. Implementing comprehensive workforce retraining programs can ensure that individuals are equipped for the jobs of the future. By investing in reskilling and upskilling initiatives, the US can mitigate the negative effects of automation and promote inclusive economic growth.

The United States has the chance to spearhead international cooperation and set global standards for responsible GenAI development and implementation. Working with other nations, sharing best practices, and establishing common guidelines will help to ensure that AI's power benefits all of humanity. Close collaboration with international partners allows the US to help shape the ethical, legal, and policy frameworks governing GenAI, cultivating a global environment of trust and conscientious innovation.

As the United States remains at the forefront of GenAI development, it is essential to adhere to the constitutional principles of liberty, equality, and pluralism. The cybersecurity community plays a crucial role in ensuring the safe emergence of AGI. By promoting public-private partnerships, attracting diverse minds, and leading in ethics, the US can champion beneficial improvements and make sure that GenAI's potential is harnessed to empower people, rather than enable oppression or sophisticated cyber threats.

While there may be confusion regarding the competitive landscape of AI development, it is important to note that both the United States and China have made great breakthroughs in the field. However, the US maintains a strong position in GenAI innovation due to its robust tech ecosystem, access to top talent, and history of pioneering research. By addressing potential social and economic challenges, fostering international cooperation, and upholding core values, the United States can continue to lead in the AI arena and ensure the responsible growth of this emerging technology.

If the development of GenAI is transparent, subject to civil oversight, and embraced with collective responsibility, it has the power to unlock vast knowledge, creativity, and human potential. However, without proper safeguards, there is also a risk of calamity. Given that the United States has successfully navigated previous technological revolutions while upholding human rights and democratic norms, we must use the same moral conviction to guide GenAI toward a future that uplifts humanity.

The ongoing AI race is a pivotal milestone for technology, and particularly the future of cybersecurity. As AI systems become more capable of flexibly responding to novel threats, they will reshape approaches to security operations and defense. One vital question therefore presents itself: how will cybersecurity businesses harness emergent AI to redefine their offerings?

Managed security service providers (MSSPs), who specialize in security monitoring, threat detection, and incident response services, find themselves at the forefront of this evolution. Positioned to rapidly enhance their capabilities through AI advancements, MSSPs face both opportunities and challenges as intelligent systems become more prevalent.

The rise of machine cognition poses a critical juncture for the cybersecurity industry, requiring MSSPs to adapt to the opportunities presented by increasingly intelligent systems while navigating associated risks. The next chapter will delve into how MSSPs are set to leverage progress to enhance detection, automation, and decision support. By exploring the factors that shape the future trajectory of MSSPs in the age of thinking machines, we will navigate the evolving landscape of cybersecurity and the role MSSPs play within it.

Chapter 20

THE FUTURE OF MSSPS IN AMERICA'S CYBERSECURITY ENDGAME

As the cybersecurity landscape undergoes rapid transformations, managed security service providers (MSSPs) in America are racing against time to adapt and stay ahead of the curve. The integration of advanced technologies like generative artificial intelligence (GenAI) is revolutionizing the role of MSSPs in safeguarding organizations' digital assets. This chapter explores how GenAI is shaping their future and their capacity to provide robust security services in the face of mounting cyber threats.

MSSPs have become crucial partners in America's cybersecurity ecosystem, offering outsourced expertise and capabilities to protect against evolving risks. The value of MSSPs lies in their ability to provide high-quality security services without requiring organizations to maintain in-house teams and infrastructure. As cyberattacks become more frequent and complex, MSSPs have taken on a strategic role, offering proactive risk identification and mitigation and playing a vital part in keeping the nation's digital landscape safe.

The emergence of GenAI presents significant opportunities for MSSPs to enhance their capabilities, accelerate threat detection, and improve operational efficiency. By integrating AI technologies such as pattern recognition and predictive analysis, MSSPs can better detect stealthy attacks, identify anomalies across diverse data sources, and anticipate novel attack vectors. In terms of America's cybersecurity endgame, the race against time necessitates that MSSPs expand their repertoire in several critical areas to effectively counter the growing risks.

A key challenge lies in detecting dangers that traditional tools may miss. GenAI empowers MSSPs to analyze data from network traffic, endpoint signals, and threat intelligence, using pattern recognition to uncover subtle anomalies and indicators of malicious activity. Furthermore, predictive capabilities enable MSSPs to stay ahead of threats by anticipating adversary behaviors, which is particularly crucial for the US's cybersecurity future.

The rise of cloud platforms like AWS and Azure demands ironclad cloud security services from MSSPs. GenAI enables MSSPs to safeguard these environments by automating the discovery of misconfigured resources and embedding security into CI/CD pipelines. This ensures the protection of containerized microservices throughout their lifecycle and supports the security of microservices architectures, contributing to America's overall resilience.

By leveraging natural language processing and other AI techniques, MSSPs can streamline their operations and optimize analyst efficiency. This permits the automation of low-value tasks, such as alert reviews, and facilitates the prioritization of threats, the documentation of incidents, and the orchestration of response workflows. Consequently, analysts can focus their efforts on high-value investigations and strategic initiatives that require human intelligence and creativity.

With the nation facing a critical cybersecurity endgame, the role of MSSPs will be defined by their ability to adapt and innovate in the face of emerging dangers. Harnessing the power of GenAI and expanding their skills will allow MSSPs to enhance their offerings and solidify their position as trusted partners, defending organizations against the ever-growing array of cyber threats.

MSSPs are a central part of shaping America's digital future. As the cybersecurity landscape evolves, concerns have arisen regarding AI's potential to replace skilled professionals. Responsible MSSPs recognize that technology should complement human capabilities rather than replace them. Their goal is to minimize mundane tasks, enabling security teams to focus on higher-impact work. Transparency in AI's capabilities and limitations will be vital for establishing trust between MSSPs and their clients.

In the context of critical incidents, which encompass a range of cybersecurity events such as data breaches, malware infections, or denial-of-service attacks, the speed and precision of GenAI systems are expected to surpass human capabilities in analyzing the scope of attacks and initiating countermeasures. While human analysts may take hours or even days to identify root causes and respond,

GenAI will be able to accomplish this within minutes. By automating containment actions, like quarantining infected hosts, response time will be significantly shortened, limiting adversaries' impact.

Certain actions, like network segmentation, will still require human judgment, so MSSPs will need to discern when to incorporate human analysts into automated incident response processes. Regardless, GenAI's accuracy and speed will enable a rapid and effective response, reducing recovery time and associated costs.

Due to escalating cyber risks, the demand for managed security services is anticipated to outpace the supply of skilled analysts. GenAI will provide an avenue for MSSPs to keep up with customer growth without compromising service quality or effectiveness. Through leveraging predictive analytics, MSSPs will be able to forecast resourcing needs and expand their capabilities to accommodate their growing clientele.

MSSP leaders will need to continually evaluate and enhance the fairness and accuracy of GenAI to avoid widespread harm resulting from inaccurate outputs. Ensuring that training datasets are unbiased and representative will be crucial. Monitoring metrics like false positives and false negatives will enable providers to gauge real-world performance beyond controlled settings. Prioritizing security and ethics will be critical for responsibly expanding GenAI capabilities.

As MSSPs incorporate AI into their operations, they will need to navigate changing regulations concerning data privacy, algorithmic accountability, and transparency.

Regulations like GDPR will require them to evaluate how they collect, process, and secure customer data used to train AI algorithms. MSSPs will also need visibility into the factors driving GenAI outputs to avoid non-compliant decisions and address "right to explanation" requirements.

MSSPs who invest in AI audits, ethics boards, and tools for tracking data lineage and explainability will be better positioned to unlock the full potential of GenAI while aligning with key regulations. A principled approach will also foster trust in the services offered by GenAI-enabled providers.

Looking forward, GenAI will likely continue to advance alongside other emerging technologies such as quantum computing, expanded connectivity from 5G and beyond, and the Internet of Things. As the attack surface broadens, maintaining effective cybersecurity measures with internal security teams alone will become impractical, increasing the importance of advanced MSSPs.

We can also anticipate further consolidation among MSSPs, with innovation and delivery models becoming critical differentiators. Specialized MSSPs catering to specific industries and offering tailored GenAI solutions for unique use cases are likely to emerge. Additionally, regulatory scrutiny concerning security, privacy, and algorithmic transparency will continue to intensify. In essence, GenAI-powered MSSPs have the potential to automate prevention and response efforts, predict emerging threats, and become essential catalysts for cyber resilience in the ever-evolving digital landscape.

Currently, MSSPs typically charge fixed monthly fees based on the number of assets or users protected. However, forward-thinking providers are expected to adopt more dynamic pricing models that align with each customer's unique risk exposure and activity levels. For instance, future MSSPs could offer pricing tiers based on guaranteed response times for on-demand incident response, considering the rising costs of breaches. More customized and usage-based models will likely emerge to cater to diverse budgets and organizational needs.

In the past, MSSPs focused on protecting individual point products like firewalls in isolation. However, modern security environments demand seamless integration and orchestration across fragmented security tools. Leading MSSPs will continue to leverage APIs and standards like OpenC2 to integrate different vendor solutions into a unified platform, customized for clients' hybrid cloud or on-premises environments. This will enable synergized prevention, detection, and response across the entire security stack.

Prominent MSSPs demonstrate their commitment to innovation by investing in research and development, recruiting top talent, and swiftly transforming ideas into advanced capabilities. This trend is expected to continue, with MSSPs actively participating in industry groups for sharing threat intelligence, coordinating with vendors on vulnerability disclosures, and engaging with the research community. A culture that values innovation and collaboration will remain crucial for an MSSP's ability to stay ahead of evolving risks.

As technology expands into new frontiers like the Internet of Things (IoT), edge devices, and decentralized networks, MSSPs will need to adapt and address threats targeting these new attack surfaces. GenAI will enable capabilities like device profiling and anomaly detection tailored to specific operational contexts, from smart factories to connected vehicles. MSSPs who specialize in securing operational technology (OT), IoT, and other specialized environments will be well-positioned to provide the necessary protection. As new use cases emerge, GenAI will facilitate the rapid development and deployment of tailored security solutions.

The shortage of skilled cybersecurity professionals poses a significant challenge for enterprises looking to build robust in-house teams. According to an (ISC)2 study, the global cybersecurity workforce gap could reach 3.4 million unfilled positions within the next few years. MSSPs that leverage the scalability of GenAI will be able to mitigate this skill deficit. Additionally, they will create opportunities for apprenticeships and training programs that bring more diversity and talent into the cybersecurity profession. By developing cybersecurity skills and partnering with MSSPs, organizations can maintain resilient security postures, even with limited internal resources.

Addressing the cyber education gap will require a concerted effort from various stakeholders, including governments, educational institutions, and businesses. Strategic partnerships between MSSPs, academia, and industry leaders will be essential to bridge this gap and strengthen the cybersecurity workforce. Together, these entities can create training programs, develop certification standards, and establish apprenticeships that equip aspiring professionals with the necessary skills to succeed in this rapidly evolving field.

As the stakes of America's cybersecurity endgame continue to rise, MSSPs must remain agile and vigilant in their efforts to protect the nation's digital infrastructure. The fusion of advanced technologies like GenAI and human expertise will be the cornerstone of effective cybersecurity strategies. MSSPs that successfully navigate the winding road ahead will be indispensable in safeguarding America's economic prosperity and national security.

The role of MSSPs extends beyond providing technical solutions; they are instrumental in fostering a culture of cybersecurity awareness and preparedness. By offering education, training, and guidance to organizations, they empower businesses and individuals to take proactive measures in protecting their digital assets. This collaborative approach ensures that organizations of all sizes can access the necessary resources to outsmart emerging dangers.

Furthermore, MSSPs are at the forefront of driving innovation in cybersecurity technologies and practices. As they tackle the challenges that our future presents, they will continue to invest in cutting-edge research, develop novel solutions, and refine their service delivery models. The lessons learned from these efforts will not only benefit their clients but also contribute to the broader advancement of cybersecurity knowledge and best practices, both within the United States and globally.

In conclusion, America's cybersecurity endgame represents a pivotal moment in the fight against cyber threats. The rapid evolution of the threat landscape demands that MSSPs evolve and adapt to maintain their position as trusted partners in securing the nation's digital future. By embracing advanced technologies like GenAI, expanding their capabilities, and cultivating collaboration, MSSPs will continue to play an essential role in protecting America's digital assets and fortifying its cyber resilience. As the nation faces the complexities of the cybersecurity endgame, the contributions of MSSPs will undoubtedly shape the path of cybersecurity in America and beyond.

CONCLUSION

As we end of our exploration of America's cybersecurity landscape, it is evident that the nation stands at a critical juncture. The race against time, a theme that resonates throughout this book, emphasizes the urgent need for collective action to fortify our digital defenses and secure a prosperous future. The title, "The Cybersecurity Endgame: America's Race Against Time," encapsulates the gravity of this moment, as the consequences of inaction could be devastating.

From the crippling ransomware attacks on critical infrastructure to state-sponsored espionage campaigns targeting sensitive systems, the threats we face are not only diverse but also rapidly evolving. The statistics are alarming, with cybercrime projected to cause trillions of dollars in damages annually. In this perilous landscape, no organization or individual is immune to the risks. The urgency of the situation cannot be overstated, as every moment counts in the ongoing battle for cybersecurity.

Throughout this book, we have stressed the importance of recognizing that cybercriminals do not discriminate based on the size of an organization when launching their attacks. While large corporations often make headlines when they fall victim to cyber threats, small businesses are equally, if not more, vulnerable to them. Small businesses are the backbone of many economies, and their security should not be overlooked. Cybercriminals frequently target them due to their potentially weaker security measures, exploiting them as stepping stones to gain access to larger organizations within the supply chain. As a result, it is essential that small businesses prioritize cybersecurity and invest in appropriate protective measures to safeguard their valuable data and assets.

This book has also underscored the pressing need for collaboration between the public and private sectors, comprehensive frameworks such as the NIST Cybersecurity Framework, and fostering a culture of cyber awareness among organizations and citizens. Furthermore, we have explored promising technologies like artificial intelligence and quantum encryption as potential game-changers in the fight against cyber threats.

In conclusion, "The Cybersecurity Endgame: America's Race Against Time" serves as a rallying cry, urging individuals, organizations, and the nation as a whole to recognize the gravity of the situation and unite in the pursuit of a safer, more resilient digital future. The clock is ticking, and the time to act is now. By understanding the risks faced by organizations of all sizes and implementing effective security measures, we can help create a more secure digital landscape and protect against the ever-evolving threats posed by cybercriminals. Together, we can turn the tide in America's cybersecurity endgame and ensure a brighter, more secure future for all.

Bonus Content

THE URGENT ROLE OF BLOCKCHAIN IN AMERICA'S CYBERSECURITY STRATEGY

As the United States grapples with escalating cybersecurity threats, innovative technologies like blockchain offer promising solutions to safeguard digital assets and critical infrastructure. Blockchain technology, a revolutionary concept that emerged from the digital age, has significantly transformed how we perceive data security, financial transactions, and digital trust. Its origins can be traced back to the early 1990s, when researchers Stuart Haber and W. Scott Stornetta introduced a computationally practical solution for time-stamping digital documents so that they could not be backdated or tampered with. This concept laid the groundwork for what would become the blockchain.

The blockchain's most notable milestone came in 2008 with the publication of a whitepaper by an individual or group under the pseudonym Satoshi Nakamoto. The paper, titled "Bitcoin: A Peer-to-Peer Electronic Cash System," proposed a decentralized digital currency system that utilizes blockchain technology to record transactions across a network of computers. This ensures that each transaction is secure, transparent, and irreversible without the need for a central authority, such as a bank.

How Blockchain Works with Bitcoin

Bitcoin operates on a blockchain network, where each block contains a number of transactions. Every time a transaction occurs, it is broadcasted to the network and verified by participants called miners. Miners use powerful computers to solve complex mathematical problems that validate transactions and add them to the blockchain. This process, known as proof of work, helps prevent fraud and ensures the integrity of the transaction record.

The decentralized nature of the blockchain means that no single entity has control over the entire Bitcoin network. This democratizes financial transactions, allowing users to send and receive funds anywhere in the world without interference from governments or financial institutions. It also provides a level of transparency, as all transactions are recorded on the blockchain and can be viewed by anyone.

Cybercriminals and Blockchain

Despite its benefits, blockchain technology has also attracted the attention of cybercriminals. The anonymity and security offered by cryptocurrencies like Bitcoin have made them a preferred method for conducting illegal transactions online, including money laundering, drug trafficking, and ransomware payments. Cybercriminals exploit the decentralized nature of the blockchain to operate outside the jurisdiction of traditional law enforcement.

Moreover, the immutable nature of blockchain transactions means that once funds are transferred, reversing the transaction is nearly impossible. This characteristic has been exploited in various scams and frauds, where unsuspecting victims send cryptocurrencies to fraudulent actors, only to find that there is no way to recover their lost funds.

Despite these challenges, blockchain technology holds significant potential for positive applications in the United States. Its capabilities can be harnessed to improve various sectors, including healthcare, supply chain management, voting systems, and more.

1. **Healthcare**: Blockchain can safeguard patient records and ensure their privacy while providing a transparent and immutable history of medical treatments. This can improve patient care, reduce fraudulent activities, and streamline administrative processes.
2. **Supply Chain Management**: By creating a clear, unalterable record of products from origin to consumer, blockchain can help combat counterfeit goods, verify compliance with ethical sourcing standards, and improve efficiency in logistics.
3. **Voting Systems:** Blockchain could revolutionize voting systems, making them more secure, transparent, and accessible. By recording votes on a blockchain, we could minimize voter fraud, ensure the integrity of elections, and simplify the voting process to boost voter turnout.

4. **Financial Services:** Beyond cryptocurrencies, blockchain can offer more inclusive financial services, reduce fees for currency exchanges and remittances, and provide safe, real-time transactions across the globe.
5. **Identity Verification**: Blockchain's secure and immutable record-keeping could offer a solution to identity theft and fraud. Using blockchain-based digital identities, individuals can have more control over their personal information, deciding who can access it and for what purpose.

America has a critical role in adopting and promoting blockchain technology. Given the urgency of defending against cybersecurity threats, we must prioritize the integration of blockchain into our strategy, focusing on protecting critical infrastructure and sensitive data. Our nation's leadership in this domain can inspire other countries to follow suit and contribute to a safer digital world.

Since its inception, blockchain technology has grown from a theoretical concept to a foundational technology behind cryptocurrencies like Bitcoin. While it has been leveraged by cybercriminals for nefarious purposes, its potential applications for societal benefit are vast and varied. By facing the challenges and focusing on the positive applications, the United States can harness blockchain technology to improve efficiency, security, and transparency across numerous sectors, making it a force for good in the digital age.

As the United States navigates the rapidly evolving digital landscape, embracing blockchain in its cybersecurity strategy is crucial to safeguard critical infrastructure, protect sensitive data, and maintain technological leadership. By understanding blockchain's key features, addressing its challenges, and exploring its potential applications, we can harness this technology's power to create a safer and more secure future.

GLOSSARY

Here's an expanded glossary of fifty top cybersecurity terms, organized alphabetically, to provide a broad understanding of key concepts, threats, tools, and practices in the field of cybersecurity.

Access Control: The selective restriction of access to a place or resource, allowing users to access only the information they are authorized for.

Adware: Unwanted software designed to throw advertisements up on your screen, most often within a web browser.

Antivirus Software: A program that detects, prevents, and takes action to disarm or remove malicious software programs, such as viruses and worms.

Application Security: Protective measures and protocols put in place to safeguard software applications from threats that can come through flaws in their design, development, deployment, upgrade, or maintenance.

APT (Advanced Persistent Threat): A prolonged and targeted cyberattack in which an intruder gains access to a network and remains undetected for an extended period.

Authentication: The process of verifying the identity of a user or process.

Backup: A copy of data created and stored elsewhere so that it may be used to restore the original after a data loss event.

Biometric Authentication: A security process that relies on the unique biological characteristics of an individual to verify that he is who he says he is.

Bot: A software application that runs automated tasks (scripts) over the internet.

Botnet: A network of private computers infected with malicious software and controlled as a group without the owners' knowledge, e.g., to send spam.

Brute Force Attack: A trial-and-error method used to obtain information such as a user password or personal identification number (PIN).

Cloud Security: A set of policies, controls, procedures, and technologies that work together to protect cloud-based systems, data, and infrastructure.

Cryptocurrency Security: Measures and technologies used to protect digital currencies and their transactions from unauthorized access, fraud, and other security threats.

Cryptography: The study of secure communications techniques that allow only the sender and intended recipient of a message to view its contents.

Cyber Espionage: The act or practice of using digital tools to obtain secret information without the permission of the information's holder.

Cybersecurity: The practice of protecting systems, networks, and programs from digital attacks.

DDoS (Distributed Denial of Service) Attack: An attempt to make an online service unavailable by overwhelming it with traffic from multiple sources.

Deepfake: Synthetic media in which a person in an existing image or video is replaced with someone else's likeness.

Digital Signature: A mathematical scheme for verifying the authenticity of digital messages or documents.

DNS Spoofing: Unauthorized alteration of DNS (Domain Name System) entries to redirect users to fraudulent sites intending to steal information or distribute malware.

Encryption: The process of converting information or data into a code, especially to prevent unauthorized access.

Endpoint Detection and Response (EDR): A cybersecurity technology that addresses the need for continuous monitoring and response to advanced threats.

Firewall: A network security device that monitors incoming and outgoing network traffic and decides whether to allow or block specific traffic based on a defined set of security rules.

Hacker: An individual who uses computers, networking, or other skills to overcome a technical problem.

IAM (Identity and Access Management): A framework of policies and technologies for ensuring that the proper people in an enterprise have the appropriate access to technology resources.

Incident Response: The approach taken by an organization to prepare for, detect, investigate, and recover from a cybersecurity breach.

IoT Security (Internet of Things Security): The technology and processes used to protect connected devices and networks in the Internet of Things (IoT).

IP Spoofing: A technique used to gain unauthorized access to computers, wherein the intruder sends messages to a computer with an IP address indicating that the message is coming from a trusted host.

Malware: Software intentionally designed to cause damage to a computer, server, client, or computer network.

Man-in-the-Middle Attack: An attack where the perpetrator secretly relays and possibly alters the communication between two parties who believe they are directly communicating with each other.

Multi-Factor Authentication (MFA): A security system that requires more than one method of authentication from independent categories of credentials to verify the user's identity for a login or other transaction.

Network Security: The practice of preventing and protecting against unauthorized intrusion into corporate networks.

Patch Management: The process of distributing and applying updates to software. These patches are often necessary to correct errors (known as "vulnerabilities" or "bugs") in the software.

Penetration Testing: A simulated cyberattack against your computer system to check for exploitable vulnerabilities.

Pharming: A cyberattack intended to redirect a website's traffic to another, fraudulent site.

Phishing: A fraudulent attempt to obtain sensitive information—such as usernames, passwords, and credit card details—by disguising oneself as a trustworthy entity in an electronic communication.

Ransomware: A type of malicious software designed to block access to a computer system until a sum of money is paid.

Risk Assessment: The process of identifying, analyzing, and evaluating risk.

Rootkit: A set of software tools that enable an unauthorized user to gain control of a computer system without being detected.

Secure Socket Layer (SSL): The standard security technology for establishing an encrypted link between a web server and a browser, ensuring that all data passed between them remains private.

Security Information and Event Management (SIEM): A set of tools and services offering a holistic view of an organization's information security.

Social Engineering: The use of deception to manipulate individuals into divulging confidential or personal information that may be used for fraudulent purposes.

Software Update: The process of installing files that will update a computer program to fix bugs or provide enhancements.

Spear Phishing: An email or electronic communications scam targeted towards a specific individual, organization, or business.

Spyware: Software that enables a user to obtain information about another's computer activities by covertly transmitting data from their hard drive.

Threat Intelligence: Evidence-based knowledge, including context, mechanisms, indicators, implications, and actionable advice, about an existing or emerging menace to assets.

Trojan Horse: A type of malicious code or software that looks legitimate but can take control of your computer.

Vulnerability: A weakness that can be exploited by a threat actor, such as an attacker, to perform unauthorized actions within a computer system.

VPN (Virtual Private Network): A network that is constructed by using public wires—usually the Internet—to connect to a private network, such as a company's internal network, enhancing security and privacy.

Zero-Day Exploit: A cyberattack that occurs on the same day a weakness is discovered in software, before the software creators have the opportunity to create a patch to fix the vulnerability.

This glossary represents a comprehensive overview of the diverse and complex terminology found within the field of cybersecurity, providing a foundation for understanding and further exploration.

Made in the USA
Las Vegas, NV
04 November 2026

57423055R00174